# Culture, Conservation and Biodiversity

## The Social Dimension of Linking Local Level Development and Conservation through Protected Areas

the
UNIVERSITY
of
GREENWICH

JOHN WILEY & SONS
Chichester • New York • Brisbane • Toronto • Singapore

Copyright © 1996 by the authors
Published 1996 by John Wiley & Sons Ltd,
Baffins Lane, Chichester,
West Sussex PO19 1UD, England

National       01243 779777
International  (+44) 1243 779777

All rights reserved.

No part of this book may be reproduced by any means,
or transmitted, or translated into a machine language
without the written permission of the publisher.

*Other Wiley Editorial Offices*

John Wiley & Sons, Inc., 605 Third Avenue,
New York, NY 10158-0012, USA

Jacaranda Wiley Ltd, 33 Park Road, Milton,
Queensland 4064, Australia

John Wiley & Sons (Canada) Ltd, 22 Worcester Road,
Rexdale, Ontario M9W 1L1, Canada

John Wiley & Sons (SEA) Pte Ltd, 2 Clementi Loop #02-01,
Jin Xing Distripark, Singapore 0512

**Library of Congress Cataloging-in-Publication Data**
Furze, Brian, 1957-
    Culture, conservation, and biodiversity : the social dimension of linking local level development and conservation through protected areas / Brian Furze, Terry De Lacy, and Jim Birckhead, with Philip Tracey and Gabrielle Wiltshire ; foreword by Jeffrey A. McNeely.
        p.    cm.
    Includes bibliographical references and index.
    ISBN 0-471-94902-7
    1. Man — influence on nature — Case studies. 2. Natural areas — Management — Case studies. 3. Nature conservation — Citizen participation — Case studies. 4. Rural development — Case studies. 5. Sustainable development — Case studies. 6. Environmental protection — Case studies. I. De Lacy, Terry. II. Birckhead, Jim. III. Title.
GF75.F87 1996
333.7' 15 — dc20                                                                95-42636
                                                                                    CIP

**British Library Cataloguing in Publication Data**

A catalogue record for this book is available from the British Library

ISBN 0-471-94902-7

Typeset in 10/12pt Sabon from authors' disks by Saxon Graphics Ltd, Derby
Printed and bound in Great Britain by Biddles Ltd, Guildford and King's Lynn
This book is printed on acid-free paper responsibly manufactured from sustainable forestation, for which at least two trees are planted for each one used for paper production.

# Contents

| | |
|---|---|
| About the Authors | vii |
| Foreword by Jeffrey A. McNeely | ix |
| Introduction | xiii |

| | | |
|---|---|---|
| SECTION 1 | LOCAL DEVELOPMENT, BIODIVERSITY CONSERVATION AND PROTECTED AREAS | 1 |
| 1 | Frameworks for Understanding Conservation and Development through Protected Areas | 3 |
| 2 | Conserving Biodiversity, Protected Areas and Local Development | 16 |
| SECTION 2 | SOCIAL SCIENCE UNDERSTANDING: PRINCIPLES AND PROCESSES | 29 |
| 3 | Using Knowledge from the Social Sciences | 31 |
| 4 | Using Methods from the Social Sciences | 48 |
| 5 | Institution Building and Community Consultation | 93 |
| SECTION 3 | ISSUES IN LINKING DEVELOPMENT AND CONSERVATION THROUGH PROTECTED AREAS | 109 |
| 6 | Rural Development | 111 |
| 7 | Indigenous People | 126 |
| 8 | Ecotourism | 146 |

| | |
|---|---:|
| **SECTION 4  MODELS OF MANAGEMENT** | 177 |
| 9  Local Level Management of Resources | 179 |
| 10  Biosphere Reserves | 209 |
| | |
| A Final Note | 247 |
| References | 249 |
| Index | 263 |

# About the Authors

**Brian Furze** is a rural and environmental sociologist, the coordinator of the Rural and Regional Development Group (RRDG) at La Trobe University's Albury/Wodonga campus and a member of the Johnstone Centre of Parks, Recreation and Heritage of Charles Sturt University. He has worked extensively in Australia and the Asian region in the areas of community-based conservation and rural development, and is the author of a number of books and papers on the subject.

**Terry De Lacy** is professor of protected area management and director of the Johnstone Centre of Parks, Recreation and Heritage at Charles Sturt University. Terry's research area is in environmental policy analysis focusing on protected areas. Recent projects include 'Aboriginal involvement in protected areas', 'Economic valuation of protected areas' and 'Landcare evaluation in Australia'. He is presently leading a team undertaking a 'Comparative study in Asia and the Pacific on linking development and conservation through biosphere reserves'. Terry is an Australian delegate to IUCN's International Commission on National Parks and Protected Areas. Professor De Lacy takes up the position of dean, Faculty of Land and Food Systems, at Queensland University later in 1996.

**Jim Birckhead** is a senior lecturer in anthropology and member of the Johnstone Centre at Charles Sturt University. Jim's research interests include Aboriginal land management practices, contemporary Aboriginality and ranger training programmes, and issues of cultural representation – Aboriginal Australia and Appalachia, ethnographic evaluation of educational and training programmes, minority identity and the media, religion and belief systems in modern societies, and cultural and racial politics.

**Phillip Tracey** was a research officer in the Johnstone Centre and is presently completing a PhD on ecotourism at the University of Tasmania.

**Gabrielle Wiltshire** has recently completed her degree in parks, recreation and heritage and is working as a research officer in the Johnstone Centre.

# Foreword

'Pristine nature' is a fine myth, carrying with it satisfying images of an idealised natural area which has not yet been disturbed by the pernicious effects of humanity. Conservationists have sought to convert this myth into reality by establishing so-called 'wilderness areas'. By 1993, the *United Nations List of National Parks and Protected Areas* included some 9832 sites covering 9,263,496 square kilometres, equivalent to the total land area of the USA. About half of this area is in the form of wilderness areas, strict nature reserves and national parks and monuments, where human influence is supposed to be minimal. But recent advances in ecology, landscape history and the social sciences have clearly indicated that nearly all the landscapes we see today in all parts of the world have been profoundly influenced by human activity in the past. Thus even the most 'natural-appearing' habitats are in fact *cultural habitats*, largely created by human influence. And seeking to maintain these areas in their current state, with their current mix of species, is also a cultural response – a purposeful intervention by people to maintain something they value.

This might sound radical, even threatening, to some protected area managers who have traditionally defined their main task as protecting nature and wilderness against the depredations of humans. And these ideas certainly imply serious new challenges to protected area managers who already feel handicapped by lack of budget, insufficient support from government, shortage of well-trained personnel, and absence of support from people in surrounding regions. But at the same time, the perspective of protected areas as cultural artifacts also provides some interesting new opportunities for managing these areas to maintain the values which are appreciated by the public. If the habitats were created by past human activity, which current human activities will serve to maintain the habitat values, and which would tend to lead to the degradation of the very features for which the area is being protected?

This book discusses these issues in considerable detail, providing new insights into how the modern protected area manager can expand his or her perspectives about the people who live in and around the sites being protected. In many cases, visitors to such sites may find that cultural attractions are often just as interesting as the natural attractions, or even more interesting. Seeing the way that people live among nature without destroying it can be an ennobling experience for urban-dwelling tourists, whose only brush with natural habitats is through the Discovery Channel, *National Geographic* or BBC wildlife television programmes.

In the recent past, protected area management was seen as primarily a biological challenge, and biologists certainly have been the leaders in the global environmental movement. But the new approach, as exemplified by this book, recognises that the management of wildlife and habitats is only a small part of the picture; the far more challenging management task is finding the means to ensure that humans behave in ways that are consistent with conservation objectives. This requires that the social sciences play a much more prominent role in the training of protected area managers, and that the insights offered by economics, psychology, history, anthropology, political science and sociology be harnessed for the benefit of conservation. The modern protected area manager needs to be a sort of renaissance person, with a solid grounding in both biological and social sciences, as well as resource management and practical diplomacy, and a sense of awe for nature and culture. Incorporating local people as legitimate partners in the management enterprise helps to convert the mission from one of protection to one of active conservation, which enables the protected area to contribute to local development objectives. The hope is that using such approaches to demonstrate the benefits of protected areas to society will help build stronger support – locally, nationally and internationally.

One key element is reaching agreement among different groups of people as to their objectives for any particular area of land. Should a national park be maintained for conserving biodiversity, maintaining watersheds or promoting tourism? Or should the area be logged and converted to agriculture which would feed growing urban populations? It is often easy to come down on the side of conservation – Uluru, Sagarmatha or Grand Canyon have little agricultural potential. But in other situations it is becoming increasingly urgent to choose rationally among such options. The increasing demands being made on virtually all protected areas are forcing hard decisions to be taken. Informed debate, and a process of negotiation which would ensure equitable distribution of benefits, can lead to an optimal resolution of such conflicts of interest – it is to be hoped in the favour of conservation.

'Sustainability' is a constant theme throughout this book, but sustainability in a highly dynamic social, economic, political and biological setting becomes a daunting challenge. It is best met when the individuals who are most directly affected are also fully involved in determining the objectives for which areas are to be managed. Building strong institutions that are adaptable and will enable rural communities living in and around protected areas to be increasingly self-reliant and in greater control over their lives will surely be an important step in the right direction. This book is a significant contribution to the growing application of social science principles to protected area management.

Feedback is a critical issue. How can the protected area manager best manage the site in a period of rapid social and economic change? What kind of biological and social monitoring measures need to be established to ensure that changes are perceived and incorporated into management programmes? Given the highly dynamic state of ecosystems, especially in the light of possible climate change, human-produced chemicals carried by air and water and changing economic conditions, how can the manager balance multiple demands, especially with the usual constraints of lack of budget and staff?

While this book cannot answer all of these questions, it does provide many useful insights into modern approaches to protected area management. It is especially innovative in suggesting a new and more integrated approach to protected area management that undoubtedly will become increasingly important in the future.

Jeffrey A. McNeely<br>
Chief Biodiversity Officer<br>
IUCN

# Introduction

The Fourth World Congress on Parks and Protected Areas in Caracas in 1992 chose the theme *People and Parks*. The conference report (IUCN 1993: 2) is introduced with the questions: 'How can protected areas contribute in sustainable ways to economic welfare, without detracting from the natural values for which they were established? How can local people be provided with more of the benefits of conservation, thereby becoming supporters of protected areas? How can protected areas be managed to support both biological and cultural diversity?'

A number of case study papers presented at the congress explored the link between community development and protected area management. After the congress, its secretary general, Jeffrey McNeely, approached Terry De Lacy at the Johnstone Centre to compile a book assessing the cases presented.

Based on our analysis of the Caracas papers, other recent literature and our own work, we perceived a need for a broader book looking at the link between conservation and local development through protected areas, but especially focusing on the use of social enquiry and methodology in achieving this link.

Subsequent to the Caracas Congress a number of international conferences have been held dealing with the question of community-based conservation. Three in particular have provided relevant case study material: the South Pacific Regional Environment Program's Fifth South Pacific Conference on Nature Conservation and Protected Areas held in Tonga in October 1993; the Liz Claiborne Art Ortenberg Foundation Community Based Conservation Workshop in Airlie, Virginia in October 1993; and the UNESCO Biosphere Reserve Conference held in Seville, Spain in March 1995.

We think it important to qualify our use of case study material. Some of the case studies are based on research work we have done ourselves, in China, Nepal and of course in various areas in Australia. In keeping with the original aim of the book (using secondary sources and providing a local/community development and conservation context) the majority of case studies are our interpretations of these secondary sources. We have attempted to use them to contextualise the issues or concepts we are addressing. Map 1 gives the location of the individual studies, numbered according to geography.

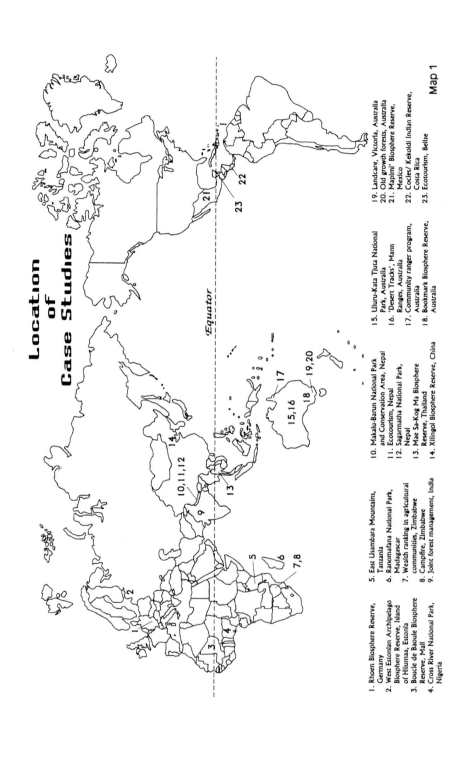

# INTRODUCTION

We have selected case studies to illustrate social science understanding and methods, to illuminate issues concerned with rural development, indigenous people and ecotourism, and to draw out principles related to models of management. We have attempted to select case studies where there are at least two case study reports or analyses to increase the reliability and validity of our presentations. We have clearly identified the information sources of our case studies and have tried to be true to their description and discussions. Nevertheless, brief summaries of often extensive reports used in very different contexts may have caused different emphasis to be given, and most importantly there was no attempt to check information or others' interpretations back in the field.

The approach we have taken is to split the book into four sections. The first, *Local development, biodiversity conservation and protected areas*, sets the scene by introducing the concepts of development and its derivatives, sustainable development, rural development, participatory development and local level development, together with biodiversity conservation and protected areas. The second section, *Social science understanding: principles and processes*, introduces material from the social sciences and describes a number of social science methods that can be used to research and understand the link between local level development and conservation. In the third section, *Issues in linking development and conservation through protected areas*, rural development, indigenous people and ecotourism are discussed in relation to their importance in understanding the conservation/development view through protected areas. Finally, the last section discusses *Models of management*, looking first at local community management of resources and then at the biosphere reserve model.

The approach therefore moves from theoretical and/or conceptual bases of understanding through social methods to the lived reality of existing projects and back again. We hope this results in a dynamic relationship between the conceptual and the practical, where each is informed by the other.

Finally, we would like to thank our many colleagues who have helped in the rather protracted development of this book. First, we must acknowledge the role of Jeffrey McNeely at IUCN who provided the catalyst for the development of the book. Also, this is an outcome of ongoing collaborative work between the Johnstone Centre at Charles Sturt University and the Rural and Regional Development Group at La Trobe University. Our sincere thanks go to our colleagues, especially Johannes Bauer, Allan Curtis, Michael Lockwood and Rik Thwaites, who have provided important insights and made many contributions. We thank Teresa Crowley and Dirk Spennemann for assisting in preparation of certain figures in the text. We also wish to acknowledge agencies who have provided research funding for projects reported in the book: Australian Research Council; Australian Department of Primary Industries and Energy; Australian Department of

Employment, Education and Training; Australian Development Aid Abroad; Chinese Academy of Sciences; Chinese MAB Committee; Nepal Department of Parks and Wildlife Conservation; Queensland Department of Technical and Further Education; and the Victorian Department of Conservation and Natural Resources. Finally, our gratitude goes to Iain Stevenson at John Wiley & Sons, who showed remarkable patience.

Brian Furze
Terry De Lacy
Jim Birckhead
*Albury/Wodonga*
July 1995

Section 1

# LOCAL DEVELOPMENT, BIODIVERSITY CONSERVATION AND PROTECTED AREAS

This section introduces you to what is and can be meant by development, local development, community development and so on. It is necessary to locate these terms and concepts within a protected area and biodiversity conservation context, because the terms and concepts are increasingly becoming part of protected area management, language and practice.

The importance of these terms to the protected area management context also highlights the importance of social science understanding. Our contention is that protected area management is not only the management of biodiversity conservation, but that it has a social (people) dimension, whether this dimension be inhabited by local people, tourists or visitors. Increasingly, this management context incorporates local people as legitimate partners in the conservation role, thus widening the role to incorporate a local development focus.

Whilst the book deals with this approach in detail, the first section locates the context within which this operates, by discussing what we mean by development and related concepts, and exploring the role which protected areas and their management have in development approaches.

# 1 Frameworks for Understanding Conservation and Development through Protected Areas

Article 8 of the *International convention on biological diversity* asserts the crucial role of protected areas in achieving the convention's objectives. However, according to Wells *et al.* (1991), many protected areas are at risk because of the hardship they place on local communities. The protection of biodiversity may therefore be seen to be one of the most pressing issues in conservation. However, it is also one of the most pressing issues in development.

With the recognition that conservation often fails to achieve its goals when local people are unsupportive, or are not meaningful partners, the question of local participation is now firmly on international conservation and sustainable development agendas. As a result, many people involved in the conservation, development and academic communities, as well as local people themselves, are involved in the search for sustainable futures.

Agenda 21, agreed to at the June 1992 United Nations Conference on Environment and Development (UNCED), emphasises the importance of involving people affected by activities in the decision-making process, of developing a consensus between local and national stakeholders and of the need to mobilise resources at all levels from the local to the global. In *Conserving the world's biological diversity*, McNeely *et al.* (1990) further highlight this and recognise the link between conservation, development and local people in protected area management. They explicitly suggest that:

> protected areas will succeed in realising their conservation objectives only to the extent that the management of the land surrounding them is compatible with the objectives of the protected areas. This will typically involve protected areas becoming parts of larger regional schemes to ensure biological and social sustainability, and to deliver appropriate benefits to the rural population. (1990: 61)

Traditional approaches to park management and enforcement activities, according to Wells *et al.* (1991: ix), have not always been able to reconcile competing environmental uses which exist between a protected area's existence and the needs of a rural populace. As a result, the biodiversity conservation role of protected areas is increasingly becoming integrated with a process of social, economic and political development based on models of ecological sustainable development.

## WHAT IS DEVELOPMENT?

Elizabeth Dowdeswell, executive director of United Nations Environment Program (UNEP), addressing the World Summit for Social Development in Copenhagen in 1995 on the confluence of the two major issues facing humanity in the twenty-first century – alleviating poverty and protecting the environment – stated:

> First, our fundamental definition of development must change. It can no longer be regarded as merely a problem of modernizing traditional societies. It should not be a mere duplication of the energy and resource-intensive development path pursued by the developed countries. It has to recognize local circumstances, potential for internally generated growth, the contribution of traditional institutions and knowledge. It has to be inherently geared towards sustainability. (Dowdeswell quoted in UNEP 1995: 9)

Development is a key term in this book. It is also a very difficult one to use, because development means different things to different people, and even can mean different things to the same people.

It is important to note that in some, perhaps many, contexts, the term development has been taken to assume a particular outcome of social action by policy makers, activists, economists, social scientists and the like who assumed that the process of modernisation which the capitalist societies of the west went through should be replicated globally. Of course, there are a series of assumptions which accompany this model, not the least being that the 'west is best', and that all societies could and should take the same path of 'development'.

Many of the people and groups who equated this idea of modernisation and industrialisation with the idea of development used 'big picture' (macro level) models to try to understand the modernisation/industrialisation process. They also looked at ways in which societies and regions which had not embarked on this process could best be modernised. Because of the global social, political and economic dominance of the modernisation model, it came to be equated with development. The term 'development' has often been used interchangeably with modernisation, and therefore has taken on a westerncentric character. It also spawned other terms which still abound and which implied similar westerncentric ideas (for example, 'undeveloped' and 'primitive').

The importance of this is that the term 'development' is actually very ideologically charged. The way it is used in popular and official dialogue suggests certain assumptions about the ways in which societies undergo social change, the 'best' or most appropriate way a society should change and, ultimately, what is best for that society. We are also entering an ideologically charged area when we talk about 'local level development' when we examine the social dimensions of linking development with protected areas.

At its barest, development could be described as the process of intervening in existing forms of society (which includes social, political and economic structures) in order to achieve desired social, political and economic goals. Development intervention occurs using a series of assumptions, including those discussed above. They may be explicit, or they may not be. You might think they are valid, others may not. Nonetheless, they are an important part of the development process. This implies that development intervention is, above all, a process based on, and subject to, power relations between competing interests.

**Plate 1.1** Continuity and change for the herding communities on the grassland of Inner Mongolia. Children outside house with satellite dish. Many similar houses exist in the area, with wind generators to provide electricity for lighting, Hi-Fi systems and television. Wind generators are also seen on more traditional yurts (felt tents). (Photo: R. Thwaites)

By acknowledging this, we heighten our awareness of the social basis of the development process. We also recognise a rich area for social conflicts originating in these power relations, exercised over the control and implementation of development agendas. The experience of rural development as well as that of many integrated conservation and development projects offers testimony to this.

So, when we talk about development, we are talking about a way in which individuals, whether they be policy makers, community development workers, economists, sociologists or, increasingly, protected area managers, actually try to intervene in social processes in order to achieve desired results. Our operational definition of development thus becomes:

> the process of intervention into existing structures of society in order to facilitate desired social, cultural, economic, political and conservation goals.

We can of course intervene in the development process at various levels and widely differing situations, whether geographic (urban, rural), sectoral (agricultural, environmental) and so on. Consequently, development has a number of groups or derivatives, which we now briefly discuss.

## RURAL DEVELOPMENT

The issue of rural development is important to our discussion. There is a long history in the social sciences related to rural development and understanding agrarian social change. Whilst this area is looked at in more detail in Chapter 6, there are a couple of important points to highlight here.

As many protected areas are located in rural areas, they are part of (or should be part of) a broader social, economic and political system of agriculture. Therefore, in a management sense, and in the specific context of linking conservation and development, many resource use issues are at least partly rural development issues.

Further, both in the broader context of development (briefly discussed above) and the specific context of local development (discussed in the next sections and throughout this book) it is important to realise that there is a range of practical and theoretical understanding on which protected area managers can draw. Throughout the book, we highlight some of the relevant experience, but, so as not to reinvent the wheel, it is useful to know that this broader experience exists. In this book, and increasingly in protected area management globally, we are talking about intervention at the local level which we call 'local level development'.

## LOCAL LEVEL DEVELOPMENT

Just what do we mean by 'local level development'? It is very easy to say, and perhaps know, that it is important to combine protected area management strategies with people who are most likely to be affected by them. But is it necessarily the case that this is a good idea and a valid strategy for local development?

Local level development is a model of development which assumes that the decision making and project implementation inherent in the process is best undertaken *by* local people (either on their own or in partnership with other agencies) and *for* local people. In other words, development becomes less of a top-down, 'we are experts and we know what is best for you' approach and more of a bottom-up, 'we are the local people, we have insights that are important to the process' approach.

It is common for local level development to be the result of a creative relationship between local people and so-called development experts. It is less common that local level development is undertaken only by local people, who are given (or who have taken) total control of the development process.

We must note the difference in the assumptions about development and the development process which local level development uses compared to other approaches. Briefly, a local level approach assumes that local people have legitimate knowledge and are legitimate partners in any development process (however development is defined). The assumption is that people have rights to an informed (by all sides) development process which takes into account rights and obligations, as well as social, economic and political characteristics.

At its best, development can provide a way in which all people are recognised for what they are: legitimate actors in social, economic and political processes which will impinge on the ways in which they live. This we can call local level development. At its worst, the development process ignores local people and assumes that they have no rights to participation. As a consequence, local people may be ignored, lose control of their lives, and see their traditional social, economic and political systems fall apart. Of course, there are also many points in between these two extremes.

Unfortunately, the local level is not without its problems. It is not uncommon that assumptions are made by development professionals about what constitutes the local level. For example, some people have assumed that the local level, like the term 'traditional', represents some romantic, better world. Others might assume that the local level consists of a number of people, all living harmoniously together (again, a romanticised view). Of course, the harsh social, economic and political reality of the local level may be quite at odds with these romanticised visions. There is much in the development experience to indicate that the local level consists of hierarchical systems of social, economic and political power.

This raises an important issue. If we are talking about local level development, are we talking about the development of all, or the development of those who have access to social power (for example, the village chief as opposed to a community meeting, the landowner as opposed to the rural labourer, men as opposed to women and so on)? If we are talking about local level development for all, then what are the implications of a local level which is hierarchical? If we only listen and pay attention to those who can speak to us (for whatever reason) have we fulfilled our goal of local level development? In other words, do we assume that because something is couched in local level terms, this will automatically result in the interests of all being represented equitably?

These are important questions which are looked at in greater detail in the following chapters. However, the central point to remember is that when we talk about local level development, the local level is not a given. All local levels are not the same, and it is not very useful to reduce the complexity of humanity's social, economic and political organisation to romanticised visions of community life. This brings us to another key point: what do we mean by community?

## WHAT IS COMMUNITY?

Perhaps one of the most often used terms in the integrated conservation development literature and experience is 'community'. A better understanding of what we mean when we talk about community is important, both to our attempts at local conservation initiatives and, in a more specific sense, as a means by which local people can become part of the development process.

There has been a long history of community studies in the social sciences, ranging in focus from analyses of rapid social change to the ways in which various community sections and groups interact to ways in which power relations are played out. As for developing a definition of what we mean when we use the term 'community', we are talking about a local social system. This definition, whilst deliberately vague, allows us to place two dimensions as important. First, we acknowledge that a community has a certain internal social, economic and political dynamic. For example, there may be senses of belonging to 'a community' (although this may not be shared), there may be geographic constraints on 'community' (where common land ends, for example), or there may be reflections of broader power relations found within 'the community' (rural inequality, for example). Secondly, we are able to recognise that communities are not self-contained systems. They reflect, and are influenced by, broader social, economic and political factors.

# UNDERSTANDING CONSERVATION AND DEVELOPMENT

There are a number of important points which emerge from this. First, there has been a long tradition in both social science (the community studies school) and development experience which has incorrectly viewed the community as an harmonious locality of self-help and relative self-sufficiency. This is what Burkey (1993) and others describe as the harmony model of rural community life. It is important to stop and think about what the implications of this can be, especially in relation to reinforcing power relations through 'development' intervention.

The fact that communities reflect broader social characteristics reinforces an awareness of rural social, economic and political relations. It is as misleading to view 'the poor' or 'the rich' as harmonious social categories as it is to view 'the community' in this way. There are different ways in which factors like class, race and ethnicity, gender and age impact on an individual's or group's experience of community life.

Secondly, in the specific context of local level participation, the fact that groups within a community will have opposing interests (even within those categories we label as 'rich' or 'poor', 'men' and 'women') means that the implementation of projects may result in conflicts.

If we acknowledge that communities are not harmonious, if we recognise that communities are influenced by broader social, economic and political processes, and if we remember that they represent an important intersection of an individual's 'lived experience' with a range of local and broader processes, then the community becomes an important vehicle for integrated conservation and development programmes. However, acknowledging these factors also highlights some of the potential difficulties and ethical considerations associated with this approach, and these are discussed in the following chapters.

## DISTINGUISHING BETWEEN COMMUNITY AND LOCAL LEVEL

In this book we do not wish to engage in long discussion over the subtle meanings of many of the terms we use. However, because concepts such as community and local level are so important to integrating development and conservation through protected areas, and they can and do mean different things to different people, some key points need to be highlighted.

As we mentioned above, 'community' is not a given, and it is not a homogenous social entity. It is full of different social groups who have differing status and power, wants and needs. When the term community is added to terms like conservation (say, for example, community conservation) it can imply that all in the community are part of the conservation programme or initiative. This can then imply a romanticised vision of community as well as an assumption that by integrating some members of a community in the initiative, we somehow have achieved our aim. Whilst the total participation of the community may be the aim, it may not be the reality, and so calling something community conservation (or community development, which has a similarly specific meaning) may misrepresent the actual project and its results. This may, at worst, lead to tokenism or, at best, result in an overstatement of the success of a project in trying to achieve community-based conservation or the integration of local people with a broad spectrum of experiences within projects and developments which affect their futures.

'Local level', we think, captures a broader meaning, but one which still incorporates the aims of community conservation and community-based conservation. Local level development may mean a number of community or community-based conservation projects integrated to become, for example, regional conservation initiatives which have their genesis at a number of 'local levels'. On the other hand, local level development may imply a more devolved conservation policy and any number of other examples which involve the participation of local people in conservation and development projects, programmes, consultations etc.

The point of the above is merely to highlight the variety of interpretations possible within the terms. It may be that projects which claim to be community conservation projects are more likely to be local level ones, because they involve only specific parts of a community or they attempt a process of devolving decision making rather than facilitating community decision making and community control in a more inclusive sense. Therefore, we should distinguish between the terms because the types of projects may have fundamentally different assumptions about ways in which the local (or community) level is incorporated into their goals.

## WHY THE LOCAL LEVEL IS IMPORTANT

Remember what was covered in the above sections on development: how we are using assumptions about desirable outcomes of the development process, as well as ways to achieve these outcomes? This section, and indeed this book, assumes that local level development is important. But what makes us assume this? And should you also assume the same?

The arguments for local level development can be arranged in two broad categories. First, there are a series of practical arguments which point to local level development being important. There is plenty of evidence to suggest that local people's integration into decision making, protected area management regimes and, more generally, the development process itself, means that there is less conflict. The ownership of the process which this integration facilitates makes for a smooth (or, at least, smoother) development process. Additionally, to ignore the local level is to ignore a rich source of knowledge. Local people have knowledge which should not be underestimated, particularly when it can be used for developing programmes such as protected area management strategies.

The second category is equally important. It relates to why, ethically, local level development is important. This position essentially states that individuals have a right to determine their own future or at least to participate in a meaningful way in decisions and processes affecting them. Given that development is a process of intervention in, and alteration to, existing social, economic and political structures, then, ethically, local people should have the right to participate meaningfully in this process.

This brings us to another key point, that of participation.

## LOCAL PARTICIPATION

It is probably reasonable to say that local participation is where local level development starts, continues and finishes. However, it is equally important to note that participation does not equal local development, nor does local development equal participation. They are mutually dependent.

We must define participation in a way that highlights the active and meaningful involvement of local people in the development process and in decisions related to it. The important point here is the term 'meaningful'. For participation to be meaningful, local involvement and consultation must mean a partnership of equals. The goal of the development process, the protected area management function and the community function must be to instigate a partnership which is based on mutual cooperation. To have local people participate in a consultation process, for example, and then dismiss their views out of hand is not meaningful participation. It does, however, set up or perpetuate antagonism between those involved in the development process.

On the other hand, if local people are consulted, and action based on mutual cooperation and a better understanding of the variety of issues involved is the result, then meaningful participation is achieved. The important result is for the parties involved in the development process to acknowledge each other's strengths and limitations, goals and wants, and construct models of development around this understanding which are meaningful to all. One major project which has taken this position from its inception is the Makalu-Barun National Park and Conservation Area Project in Nepal.

### Makalu-Barun National Park and Conservation Area Project, Nepal*

The goal of the Makalu-Barun National Park and Conservation Area Project is to protect the biodiversity of its region through the establishment of a framework and process which integrates national park management with participatory conservation area management. The project gives local communities (in partnership with the project) a greater stake in biodiversity protection. This is achieved through the use of traditional and new technologies and management capabilities for improved community development, biodiversity protection and natural resource management.

The project has integrated ecological, resource management and community development research with project development programmes, to provide conservation/development mechanisms. Each of these strands emphasises a process which is based on community involvement and community consultation.

One of the more important outcomes of the project in its early stages of development has been a series of reports, highlighting not only research which has been carried out, but also the process of project development. The project has thus been able to document information gathering, project development and application as a chart of its experience.

The overall philosophy and aims of the project (as highlighted in the community resource management document) are to:

> protect and conserve the environment with the help of the community. People are the basic resource to promote the objectives of environmental and cultural conservation through sustainable means. (Nepali *et al.* 1990: 31)

The project acknowledges that the people of the area have a high degree of integration with their natural environment, and that the project should make the most of this. Integrating this acknowledgement with a broader community development approach has resulted in: local people being considered a resource; traditional institutions being integrated into the development approach; the launching of poverty mitigation projects; the provision of appropriate skills and knowledge to local people; the creation of socially relevant programmes; and community participation being given the highest priority.

Within the frameworks of the above philosophy, a number of strategies are being put into place to: impart information, skills and knowledge to local people; initiate village level activities; promote economic activities;

---

*The development of the Makalu-Barun National Park and Conservation Area Project has been well documented. This case study draws predominantly on a community resource management report by Nepali *et al.* (1990) and primary research by the authors. See also Chapter 5.

identify potential activities; promote people's participation; and further refine an overall implementation strategy.

To ensure that the philosophy and strategies are integrated, the project has: involved communities and local institutions in all aspects of project design, development and implementation; held interactive planning meetings to encourage local people to highlight their responses and thoughts; and formed user groups to maximise local participation in planning, management, resource and economic development.

The user group concept developed in the Makalu-Barun project is an important one. Whilst organisationally it provides a mechanism whereby local people can participate in the planning and management structure of the project, philosophically it represents a mechanism whereby they can take more control over decisions which affect their everyday lives. Because user groups are encouraged, and are geographically located throughout the project, they also reflect a variety of 'local levels' and highlight the importance of not viewing local people as an homogenous group.

## SUSTAINABLE DEVELOPMENT

The World Commission on Environment and Development (1987) defined sustainable development as meeting the needs of the present generation without compromising the needs of future generations. Whilst this has been a very important term and development concept in recent years, it is not a very precise one. If it is difficult to understand what sustainable development means, it is perhaps doubly difficult to think of ways in which it is likely to be implemented.

However, it is an important dimension to contemporary conservation and development issues because it recognises a development model which, at its barest, attempts to integrate conservation with a new definition of development. Of course, not all definitions of sustainable development are going to be the same, nor is there going to be general agreement on how to implement it.

Protected area managers are therefore faced with something of a dilemma. Part of their protective function is to conserve biodiversity. However, increasingly, there are expectations that this protective function will be combined with a development function. Not only does this mean that protected area managers are faced with a fundamental shift in the management models which they are being expected to use, but part of the new model is a vague concept which has little precision and often even less operational value due to its imprecision.

## AGENTS OF CHANGE

From the above discussion it can be seen that the implementation of local development often requires some form of change agency. Local development and sustainable development may not evolve out of existing social, economic and political arrangements. What is often required is direct intervention in these arrangements in order to facilitate the local development process.

Given the emerging dual role of protected areas as both conservers of biodiversity and catalysts for local development, it is important to realise that protected areas and their management are increasingly being called upon to be agents of change. Seeing protected areas in this light results in a greater awareness of the social and conservation role that their management will play. It also brings an increased awareness of the responsibilities that this expanded role entails.

Social change of itself can be positive or negative. An understanding of not only our role as facilitators of change, but also the ethical and philosophical bases for a local level approach to conservation and development, means that we will be more likely to achieve the potentials that positive outcomes bring. These very important issues are discussed in greater length in Chapters 2, 3 and 4.

Finally, we need to understand that conflict between the competing interests of social groups is likely to arise. Tensions between not only the different groups within communities, but between differing interests external to the community, are all likely to impact on the process of local development. Conflict resolution, therefore, is an integral part of the change agent's role. The tensions which lead to potential conflict situations are explored more fully throughout the book.

**Conflict Resolution**

Conflict can be defined as any situation where there is a clash of interests (Lewis 1993). This definition is an important one, as it highlights the varied forms of conflict (management versus local people, management versus powerful local groups, management versus non-governmental organisations and so on) and, as a consequence, the diversity of approaches to conflict resolution. Lewis has developed a number of general principles of conflict resolution which are important:

- focus on underlying interests;
- address both the procedural and substantive dimensions of the conflict;
- include all significantly affected stakeholders in arriving at a resolution to conflict;

- understand the power that stakeholders have, and take that into account when trying to resolve conflict.

What this tells us is that, as far as possible, conflict needs to be resolved within a win–win situation, where disaffection for the outcome is kept to an absolute minimum. Of course, this is easier said than done.

# 2 Conserving Biodiversity, Protected Areas and Local Development

This chapter introduces the importance of biodiversity conservation and the roles which protected areas can have in this task. A discussion of this sets the scene for us to move on to another important issue: the roles which protected areas can play in sustainable development.

## CONSERVING BIODIVERSITY

What is biodiversity, and why is it important? What are the major threats to its maintenance? What strategies are best put in place to conserve it? These issues are both complex and central to our understanding of conservation and sustainable development, and have been addressed in many recent texts. Two of these are especially relevant, *Conserving the world's biological diversity* (McNeely et al. 1990) and the *Global biodiversity strategy* (World Resources Institute et al. 1992).

### What is Biodiversity?

McNeely et al. (1990: 17) define biodiversity as 'an umbrella term for the degree of nature's variety'. It 'encompasses all species of plants, animals, and microorganisms and the ecosystems and ecological processes of which they are part'. Biodiversity can be seen as a measure of nature and its diversity, rather than an entity in itself, and is usually measured at three levels – genes, species and ecosystems (Figure 2.1).

Genetic diversity is the variety of genes within species. This form of diversity can be between populations of the same species, or within distinct populations. Species diversity is the variety of different species found in an area – the number of different species is often used as a measure, but in some cases taxonomic diversity is used, as it considers the relationship of species to each other. For example, an area is more diverse if it has species from different groups than if all the species are from the same group. Ecosystem diversity, as the name suggests, refers to the number and distrib-

# BIODIVERSITY, PROTECTED AREAS AND LOCAL DEVELOPMENT

ution of different ecosystems, a concept mostly used at the national or subnational levels. It can also refer to the diversity of habitats and processes occurring within ecosystems. Because ecosystems are not closed systems (they interact with adjoining systems) it is difficult to define them, but the assessment of biodiversity at this level is certainly very important, especially in determining priorities for conservation.

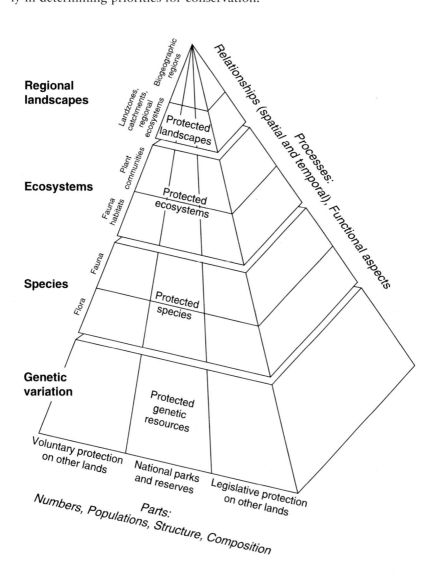

**Figure 2.1** A conceptual hierarchy of biodiversity, its parts and processes, at four levels: regional landscape, ecosystem, species and genetic variation (adapted from Sattler 1991)

In a general sense, the term 'biodiversity' is normally used to refer to having a desirable or wide range of genes, species and ecosystems, and will be used in this context in the following sections.

The schematic representation in Figure 2.1 of the different levels of biodiversity is a useful way of illustrating the different levels at which biodiversity is important. It should not, however, be taken to imply a hierarchy of importance where, for example, the bottom level is more important than the upper levels. Because biodiversity is concerned with interrelationships, all of these types of diversity are of equal importance.

**Why is Biodiversity Important?**

There are many different ethical and philosophical positions on nature and our relationship with it. Some positions, for example, would argue that nature has value beyond any that humans hold for it. Others would argue that nature is only important to the extent that it can be used for human activity. Therefore, humans relate to nature in a variety of ways.

The natural systems and processes of biodiversity are the fabric of life on the planet. They also provide the environment in which humans live. This ecological context, however, sits within a social context. We as human beings have a relationship with nature which depends on complex social, cultural, economic and political processes. Understanding the importance of biodiversity conservation, therefore, requires not only an understanding of ecological processes, but social ones as well. One commonly used categorisation of biodiversity values to humans breaks down biodiversity into ecosystem services, biological resources and social benefits.

*Ecosystem services*

Ecosystem services are broad natural systems and functions. They provide and regulate water resources, the soil, nutrient storage and cycling, pollution breakdown and absorption, and other functions such as climate stability and recovery from unpredictable events.

Humans rely heavily on these ecosystem services. Water supply is affected, for example, by whether or not catchments are vegetated, while soil health is especially important for agriculture, and the pollution breakdown and absorption capacity of natural systems is very necessary to deal with human wastes. In addition to survival needs, however, humans also use ecosystem services for non-essential purposes, specifically as a result of processes of industrialisation and development being simplistically equated with economic growth (see 'What is development?' in the previous chapter). Many conservationists are concerned that human activity has impacted upon these ecosystem services in such a way that they have now reached or exceeded the limits of their ability to support this activity.

## Biological resources

This category of biodiversity's importance emphasises its use value for human activity. Individual species provide the raw materials for many human uses. Domesticated and wild plants and animals provide the majority of food for the world's population. In addition to foods presently used, and species with potential food uses, other species are becoming increasingly important as sources of genetic diversity, which may provide resistance to disease or better productivity in agriculture. Biodiversity also provides medicinal resources – traditional medicines based on plant and animal species form the basis of health care for 80% of the world's population, in many nations, while a quarter of prescribed medicines in the USA, for example, contain active ingredients derived from plants (McNeely et al. 1990). Biological resources are also the source of many chemical compounds which are now produced synthetically, such as aspirin. Raw materials, such as woods and fibres, and ornamental blooms are supplied by many species. Some production occurs in a wild state (such as timber harvesting from naturally occurring forests) but cultivation is also important (silk production, for example).

In addition to existing use values, biodiversity can also provide for potential use values. Potential uses, such as new and useful products supplied by species, or potentially important resources (such as genetic material) that can be used in improving presently utilised species, are important reasons for biodiversity conservation. Another important consideration is that having a high level of biodiversity enhances the potential for future resources becoming available. Of the 30 to 50 million species that may actually exist on earth, less than one and a half million have been described (McNeely et al. 1990), and far fewer studied in any detail. The greater the diversity that can be conserved, the more likely we are to find useful species when needs arise.

## Social dimensions

The social dimensions of biodiversity are crucial. It is important to understand that human beings interact with the natural world and the way that this occurs is a result of a variety of ecological, historical, social, cultural, economic and political factors. We would expect, therefore, to see a variety of ways in which biodiversity conservation can be viewed in the context of social benefits.

Some groups in society (for example indigenous people as part of their belief systems) have very strong cultural attachments to species or habitats. As a result, aesthetic, inspirational, spiritual and educational needs may all depend to some extent on diverse natural systems.

The ethical and philosophical position of placing humanity as part of a broader non-human world is also very important. Such a position highlights a non-anthropocentric or non-human-centred view of all life forms and many argue that such a perspective is crucial for the protection of nature.

In a more concrete sense, nature-based recreational activities (such as fishing and wildlife photography) are pursued by many people (Driver *et al.*, 1996). Environmental features are a dominant motivation behind much international tourism. For example, the rise of ecotourism, with the specific purpose of experiencing environmental values, is evidence of the importance of biodiversity.

Finally, the knowledge that natural systems and species exist is valuable in itself for many people. This may be despite the fact that most will never experience the places and things they think are important. These existence values can be considerable, as is evidenced by the concern and support expressed to prevent rainforest destruction or species loss in many places.

**Plate 2.1** Elephants in evening light, Hwange National Park, Zimbabwe. Elephants have a high profile in the international discussion on biodiversity and extinctions. As a large animal capable of doing a great deal of damage to trees, housing and crops, they are often chased off communal lands in some African countries. With large populations restricted to protected areas, they can also do great damage within these areas, altering the local conditions from forest to grassland. (Photo: R. Thwaites)

# BIODIVERSITY, PROTECTED AREAS AND LOCAL DEVELOPMENT

## What are the Main Threats to the Conservation of Biodiversity?

The *Global biodiversity strategy* (WRI *et al.* 1992) has identified six fundamental causes of biodiversity loss. These are: the unsustainably high rate of human population growth and natural resource consumption; the steadily narrowing spectrum of traded products from agriculture, forestry and fisheries; economic systems that fail to value the environment and its resources; inequity in the ownership, management and flow of benefits from both the use and conservation of biological resources; deficiencies in knowledge and its application; and legal and institutional systems that promote unsustainable exploitation.

### Human population growth

The global population growth issue is well known to most people. The *Global biodiversity strategy* identifies the rates and magnitude of growth, and the eventual size of the global population as critical for biodiversity. Human use of the earth's resources has increased as the population has grown. At present, it is estimated that people use or destroy 39% of photosynthetic material produced on land. As human population grows, resource needs increase, and most agree that the present rates of resource use are unsustainable (WRI *et al.* 1992).

This particular emphasis on population growth and its link with resource use is not accepted by all. For example, critics of this perspective would argue that growth in resource use should be more specifically linked to poverty, industrialisation (particularly that found in the more powerful nations of the so-called 'developed' world) and the global distribution of economic benefits, food products, literacy and numeracy, amongst other issues. This highlights the sometimes simplistic way that resource use is equated with population growth. It also deflects and shifts responsibility from more powerful industrialised nations to less powerful agrarian ones.

### Narrowing spectrum of traded products

The global exchange economy that has emerged over the past century has narrowed the range of products used and traded, and made countries more dependent on each other. Agriculture has specialised in crops that will sell on world markets, leading to the decline of local species that were more suitable in traditional agricultural ecosystems. Forests with many species are converted to cash crops in order to satisfy demand for one product. Fisheries are also affected, as drift netting captures huge quantities of desired species and massive numbers of other species.

*Economic systems that fail to value the environment*

Natural environment qualities are commonly undervalued or not included at all in decision making, which is often made purely on the basis of market economics. Natural areas are converted to agricultural uses or logged even when the net effect on society is negative. This occurs because the market values of exploitation are easily recognised while conservation values are hidden or difficult to quantify. The fact that biodiversity benefits accrue to many people means that benefits are diffuse (and therefore difficult to perceive or measure) and exacerbate the problems of biodiversity being undervalued and therefore disadvantaged in the decision-making process. In many cases, property rights for areas with high biodiversity values are more likely to be obtained by those who clear or change natural areas than those who use them in a subsistence manner. Apart from the injustices inherent in this, insecure property rights can lead to exploitation of resources, as these rights may be lost at some point in the future. We return to this issue of property rights in more detail in Chapters 8 and 9.

*Inequity in the ownership, management and flow of benefits from use and conservation of resources*

Inequitable distribution of benefits and costs, based on inequitable social, economic and political structures, can lead to rapid exploitation and loss of biodiversity. This inequality can occur both within and between nations. For example, if a group in society owns or controls most of the land (or resources) they may exploit this. This may result in the rest of that society bearing the often massive costs of those actions.

Further, so-called 'developed' nations burden so-called 'less developed' nations with debt and are often responsible for exploiting the resources of these nations. As a result, the more powerful nations reap the benefits of this exploitation, while the less powerful nations incur substantial environmental, social, economic and political costs.

*Deficiencies in knowledge and its application*

The emphasis on scientific knowledge for understanding biodiversity conservation has resulted in traditional knowledge being lost or having little legitimacy in the eyes of many people. Even when knowledge is sufficient to make relatively informed decisions, it often does not reach decision makers. In addition, the relationship between the policy process and local communities may be problematic. The communication of information and decision making between policy makers who wish to conserve biodiversity, and local communities who are not aware of the benefits to them of doing so, may fail. Conversely, it may fail between local communities who are aware of benefits to them and policy makers who may support inappropriate initiatives.

*Legal and institutional systems that promote unsustainable exploitation*
Biodiversity conservation requires a cross-sectoral approach (economically and institutionally). Most national and international institutions, however, operate strictly along sectoral lines. Government and planning functions also tend to be centralised and closed, hindering local participation and obstructing access to groups within society and non-governmental organisations. Institutions with responsibility for conservation are often financially and politically disadvantaged, even when they are part of government, and they may lack coordination between them and be unable to plan strategically or comprehensively. Conservation and environmental protection laws are also lacking in many areas, and the breakdown of traditional or customary law is replaced with a culturally inappropriate legal system that allows exploitation.

Arrangements that are made to protect biodiversity tend therefore to be piecemeal and insufficient. Institutions responsible for protecting an endangered species, for example, may have no control over many aspects of the total system that affects the species' survival. A protected area may be created, but if it is affected by polluted water from outside, or by nearby clearing, it may be impossible to achieve conservation objectives.

## Strategies for Conserving Biodiversity

The *Global biodiversity strategy* (WRI *et al.* 1992: 27) details seven priority areas of action for biodiversity conservation.

1. The establishment of a national policy framework for biodiversity conservation (including reforming existing policies, adopting new policies and economic methods, and reducing demand for biological resources).
2. The creation of an international policy environment which supports national biodiversity conservation (including the incorporation of biodiversity conservation into international economic policy, strengthening the international legal framework for conservation, using development assistance to conserve biodiversity, increasing funding for biodiversity conservation and spending innovatively and effectively).
3. The creation of conditions and incentives for local biodiversity conservation (including correcting the imbalances of land and resource control that cause biodiversity loss, developing new resource management partnerships between government and local communities, expanding the sustainable use of products and services from the wild for local benefits, and ensuring that the owners of important local knowledge of resources benefit when that knowledge is used).
4. The management of biodiversity through the human environment (including creating institutional conditions that allow bioregional con-

servation, supporting private biodiversity conservation initiatives, and incorporating biodiversity conservation into the management of biological resources).
5. The strengthening of protected areas (nationally and internationally and enhancing their role in biodiversity conservation, and ensuring their sustainability).
6. The conservation of species, populations and genetic diversity (including strengthening the capacity for conserving these in natural habitats and for off-site conservation and education).
7. The expansion of human capacity to conserve biodiversity (including increasing the awareness and appreciation of biodiversity values, helping disseminate information needed to conserve biodiversity, promote basic and applied research on biodiversity conservation, and develop a human capacity for biodiversity conservation).

The conservation of biodiversity is thus a multidimensional problem. As the *Global biodiversity strategy* highlights, it is a problem which cannot be addressed without recognising the relationship between ecological, social, historical, cultural, economic and political factors on a global, regional, national and local scale. The strategy has also highlighted an important means of addressing these complex issues – that is, the role of protected areas and their formation, use and management. We now turn our attention to this role.

## PROTECTED AREAS: THEIR ROLE IN BIODIVERSITY CONSERVATION

The first two clauses of Article 8 of the International Convention on Biological Diversity on *in-situ* conservation state:

Each contracting party shall, as far as possible and as appropriate:

a) establish a system of protected areas or areas where special measures need to be taken to conserve biological diversity; and
b) develop, where necessary, guidelines for the selection, establishment and management of protected areas or areas where special measures need to be taken to conserve biological diversity.

Like many terms in general use, 'protected' means many things to many people. The notion of protection, as exemplified in the formation, use and management of protected areas, can imply both a protection from (e.g. development), as well as a protection for (e.g. biodiversity conservation, tourism and, increasingly, local development). It perhaps should not be that unusual, therefore, if we were to discover that these different uses and meanings of the idea of protection can lead to some complex and difficult tensions. We have, for example, the tension of reserves established to 'protect' from development now being charged with promoting development, particularly local level sustainable development. Obviously, the issue is not one of merely interpreting words differently. Rather, it is one of rights, responsibilities and power relations. One of the aims of this book is to understand more fully these tensions and how they have or have not been resolved through protected area formation, use and management. However, before going on to this, it is important to provide a brief discussion of protected areas, what they are and what they do.

Groups and communities have historically protected certain areas. In Australia, for example, several of the major national parks are today owned and jointly managed by Aboriginal communities. This represents a continuation of thousands of years of what we would describe as land management, such as encouraging certain activities (mosaic burning, for example) and preventing others (such as hunting in certain seasons). This protection was often enforced through complex cultural, religious, spiritual and customary practices and laws.

Chinese writing, some 3000 years ago, expressed views about nature conservation, and described regulations protecting certain areas. *Da Ju*, published in the sixth century BC by Yi Zhou Shu, observes 'do not cut down the trees during spring in order to benefit the growth of herbs. Do not fish the rivers and lakes during summer in order to benefit the growth of fish and other aquatic life' (quoted in Li 1993). An edict from the prime minister of Qi at that time, Guan Zuong, states that 'Pu mountain is a forbidden area because of the tea trees there; someone must suffer capital punishment if this law is defied' (quoted in Li 1993).

Protected areas as a land use in modern western societies evolved in the USA with the establishment of national parks at the end of the nineteenth century. The objectives of the early US park system were to protect the wildest and most beautiful (to the early colonists) areas for outdoor recreation activities. With the twentieth century being one of western and especially American domination of ideas, culture and economics, it is not surprising that the US national parks concept spread internationally.

This model, which builds a symbolic (and in some cases actual) fence around parks, removes the park from its ecological and social contexts. While local populations have been aware of the inappropriateness of this model since it was first used and exported as a resource management strategy, it has only been in recent times that this has been recognised by some sections of the international conservation community.

Currently the protected area concept, as reflected in *Parks for life: report of the IVth World Congress on National Parks and Protected Areas* (IUCN 1993), involves an integration of biodiversity conservation and sustainable development for local and regional areas. Protection, therefore, is now based on the integration of protected areas with the surrounding landscape and people.

**Classifying Protected Areas**

The world conservation union (IUCN 1994: 7) defines a protected area as:

> An area of land and/or sea especially dedicated to the protection and maintenance of biological diversity, and of natural and associated cultural resources, and managed through legal or other effective means.

All categories of protected areas should fall within this definition. But although all protected areas meet the general purposes contained in this definition, in practice the precise purposes for which protected areas are managed differ greatly. The following are the main purposes of management (IUCN 1994):

- scientific research
- wilderness protection
- preservation of species and genetic diversity
- maintenance of environmental services
- protection of specific natural and cultural features
- tourism and recreation
- education
- sustainable use of resources from natural ecosystems
- maintenance of cultural and traditional attributes.

**Protected Area Management Categories**

By way of recognising the diversity of interpretations of the protected area concept, IUCN uses six categories for classification according to the management objectives of the sites (IUCN 1994: 17–23).

I   *Strict nature reserve/wilderness area*: protected area managed mainly for science or wilderness protection.
II  *National park*: protected area managed mainly for ecosystem protection and recreation.
III *National monument*: protected area managed mainly for conservation of specific natural features.
IV  *Habitat/species management area*: protected area managed mainly for conservation through management intervention.
V   *Protected landscape/seascape*: protected area managed mainly for landscape/seascape conservation and recreation.
VI  *Managed resource protected area*: protected area managed mainly for the sustainable use of natural ecosystems.

**Matrix of Management Objectives and IUCN Protected Area Management Categories**

| Management Objective | IA | IB | II | III | IV | V | VI |
|---|---|---|---|---|---|---|---|
| Scientific research | 1 | 3 | 2 | 2 | 2 | 2 | 3 |
| Wilderness protection | 2 | 1 | 2 | 3 | 3 | – | 2 |
| Preservation of species and genetic diversity | 1 | 2 | 1 | 1 | 1 | 2 | 1 |
| Maintenance of environmental services | 2 | 1 | 1 | – | 1 | 2 | 1 |
| Protection of specific natural/cultural features | – | – | 2 | 1 | 3 | 1 | 3 |
| Tourism and recreation | – | 2 | 1 | 1 | 3 | 1 | 3 |
| Education | – | – | 2 | 2 | 2 | 2 | 3 |
| Sustainable use of resources from natural ecosystems | – | 3 | 3 | – | 2 | 2 | 1 |
| Maintenance of cultural/traditional attributes | – | – | – | – | – | 1 | 2 |

Key:
- 1    Primary objective
- 2    Secondary objective
- 3    Potentially applicable objective
- –    Not applicable

Source: IUCN (1994: 8)

## Protected Areas Designated Under International Agreements

UNESCO has established international protected area categories which recognise areas of international significance (UNESCO 1995a, Internet 1995a)

- *Biosphere reserves*: part of an international scientific programme – the UNESCO Man and the Biosphere programme. Biosphere reserves are designated for a range of objectives, including research, monitoring, training and demonstration as well as conservation, with a special emphasis on the human component.
- *World heritage sites*: declared under the Convention Concerning the Protection of the World Cultural and Natural Heritage. The areas designated have 'outstanding universal value', and the convention aims to foster international cooperation in conserving these areas.

## What are the Threats to Protected Areas

There are some 8000 protected areas in the world, covering around 750 million hectares, and accounting for 5.1% of terrestrial ecosystems (WCMC 1992). Whilst these figures would indicate a relatively substantial protected estate, a number of common and significant threats to protected area systems worldwide have been identified. Bridgewater (1992) summarises these as:

- conflicts with local people
- lack of policy commitment at nation state level to adequately protect systems
- ineffective management by trained staff of individual protected areas
- funding is insufficient or unsure
- inadequate public support

*Parks for life* (IUCN 1993) addresses these threats, and suggests an action plan to act as a framework for implementing strategies. These aim to ensure that protected areas can continue to be a mechanism for conserving biodiversity, while at the same time becoming a model for sustainable development.

These and other strategies assume certain understandings of not only local people, but also the role of protected areas in sustainable development. However, these understandings are not without difficulties, as we have already suggested and will continue to explore.

## CONCLUSION

Whilst it is difficult to set such things in concrete, there are a number of general points which emerge from the descriptions and discussions in Chapters 1 and 2. Biodiversity conservation, while important to the survival of the earth's ecosystems, needs to be understood in a context which includes social, economic and political influences as well as biological and management ones. Generally, the integration of local people into the conservation equation provides a means by which the protective function of areas can be enhanced. What has to be aimed for is a process of consultation, negotiation and participation which is based on the meaningful exploration of needs and issues. This has to be undertaken in an atmosphere of partnership and mutual trust.

Section 2

# SOCIAL SCIENCE UNDERSTANDING: PRINCIPLES AND PROCESSES

The previous section sets the scene for an integrated approach to both conservation and development. It highlights the importance of local level development, and of seeing conservation issues as development issues. When we take on this perspective, we locate people very firmly in the conservation equation.

Section 2 provides a basis for using knowledge gained from the social sciences in the search for integrating conservation and local level development. It does this by first exploring the basis for social science knowledge. It then goes on to focus on social science methods and their usefulness for protected area management and integrated approaches. Finally, the section looks at two key social science processes which are crucial to an integrated approach which is locally based – that of institution building and community consultation.

# 3 Using Knowledge from the Social Sciences

Chapters 1 and 2 introduced local level development and explored the role of protected areas in biodiversity conservation. As we have previously discussed, development is essentially a process which is social, economic and political in origin and in its operation a process which impacts directly on the environment. As a consequence, it becomes very important to know the ways in which social sciences can be used to understand the relationship between conservation and development, and its implications for protected area management. First, we need to look at what social science is.

## WHAT IS SOCIAL SCIENCE?

Put simply, social science is the body of theory, understanding and experience which deals with the human (or social) dimensions to issues. Within a social science framework, we can deal with the ways in which society operates, how it works and how its workings impact on various individuals and groups of people.

Obviously, to understand society, we need to understand the experiences of people and how society functions. Therefore, social science understanding incorporates such things as history, anthropology, cultural studies, sociology, politics, economics, psychology, communication studies and geography.

A crucial dimension to social science is the basis for its understanding. Social science deals with people and their relationships (with each other and with nature). There are no final laws which govern the ways people will or will not behave, for this is the result of complex interactions between individuals on the basis of shared or unshared cultural values. It is therefore very different to the ways of understanding through which disciplines such as physics, chemistry, ecology and so on might see the world.

So, in social science we are dealing with human diversity, not immovable iron laws.

## THE USES OF SOCIAL SCIENCE KNOWLEDGE

Social science knowledge provides an understanding of the ways in which society, social change (and, by implication, development) are the products of social forces. This can perhaps best be summed up by Giddens (1989: 654):

> As we peer over the edge of our century into the next, we cannot foresee whether the coming hundred years will be marked by peaceful social and economic development, or by the multiplication of global problems – perhaps beyond humanity's ability to solve. Unlike the early [social scientists]... we see very clearly that modern industry, technology and science are by no means wholly beneficial in their consequences. Our world is much more populous and wealthy than ever before; we have possibilities to control our destiny and shape our lives for the better, quite unimaginable to previous generations – yet the world hovers close to... ecological disaster. To say this, is not in any way to counsel an attitude of resigned despair. If there is one thing which [social science] offers us, it is a profound consciousness of the human authorship of social institutions. As human beings, aware of our achievements and limitations, we make our own history. Our understanding of the dark side of modern social change need not prevent us from sustaining a realistic and hopeful outlook towards the future.

As humans, our relationship to the natural world is by definition a social one. The ways we see nature and the ways we use nature are products of both how societies are organised and how we, as members of society, see nature's value (however value is defined). In other words, our relationship to nature is socially constructed and socially patterned.

Such a simple statement actually masks a very complex relationship between society, dominant groups, dominant social values and nature. Using protected areas as mechanisms for biodiversity conservation through local level development is actually an attempt to mediate between these factors in the nature/society relationship and to intervene in it where appropriate.

As we have noted already, systematic knowledge about human behaviour, culture and society comes to us through the social sciences. They represent a number of disciplines, approaches and truth claims, yet all are concerned with aspects of human study. It is important, therefore, for managers, policy makers and planners involved in conservation and local level development to be at least somewhat familiar with the human sciences. They are then in a better position to understand the cultural and social processes of peoples with whom they work.

We want to avoid quibbling over disciplinary boundaries and preserves. Our focus, rather, is key social science concepts, principles, methods and applications. No one will become an expert social scientist from reading these pages, but hopefully they will heighten understanding of human dimensions of a variety of social, cultural and political issues.

## The Uses of Sociology and Anthropology

Sociology has been variously described as 'the study of human groups and societies' (Giddens 1989: 732) and the 'scientific study of human society and social behaviour' (Robertson 1989: 426). It essentially deals with trying to explain the relationship between the individual and her/his society. To do this, sociologists use theoretical perspectives and concepts to try to develop explanations and further their understanding of this relationship. Sociology has historically been concerned with understanding the structures found within western industrialised society. However, there is also a long sociological tradition of considering the nature of development, particularly in agrarian societies.

Anthropology is the discipline most concerned with cultural and social dynamics at the local or small-scale level. Long associated with the study of 'exotic and bizarre' 'primitive' or tribal people in remote deserts, rainforests or South Pacific atolls and of strange cults and subcultures within first world societies, it has been seen by many as the study of the 'other', of those who are not fully western or mainstream. Using the main ethnographic fieldwork method of participant observation, anthropologists often spend years grounded in the particular, the local, getting to know life from the inside and through local eyes and languages. Ethnography quite simply means 'the process of recording and interpreting another people's way of life' through 'intimate participation in a community and observation of modes of behaviour and the organisation of social life' (Keesing 1981: 5).

Anthropology is usually associated with non-western, rural, tribal, indigenous or local peoples. Unfortunately, anthropology for many people still carries with it the stigma of its colonial past, when it assisted, or was not able to prevent, colonisation of a number of countries. Some tribal people and newly formed countries today are reluctant to deal with anthropologists as they seem to conjure up for them an image of the study of primitive people, an image many local people do not want applied to themselves. Some people who have been studied by anthropologists have also criticised the discipline for overgeneralising the features of a culture to fit into larger theoretical or evolutionary frameworks, thereby suppressing individuality or the idiosyncratic.

There are a number of broad areas of sociology and anthropology which have a direct or indirect bearing on integrated conservation and development and of which it is important to be aware. Whilst some are dealt with in quite a lot of detail throughout this book, the following are important because they set the scene for a sociological and anthropological interpretation of the issues.

*Approaches to social understanding*

In a sense anthropology and sociology complement each other. The sociological tradition has tended to focus on broader macro level processes and structures such as economic systems, systems of agricultural production, political systems and so on. On the other hand, anthropology looks more at micro level aspects of social life by focusing on individuals and the ways in which the broader social systems impact on their lives. Both approaches emphasise the importance of in-depth understanding of the complex relationship between an individual and her/his society and community.

This emphasis places anthropology and sociology in a good position to understand the complex relationships between local people, the nation-state and the global economy. The anthropological and sociological perspectives also provide considerable insight into the nature of communities, with their shifting power structures, factions, dynamics and contradictions. As discussed in Chapter 1, all too often concepts like 'community' are idealised by policymakers and assumed to exist in social and environmental harmony. Anthropological and sociological analyses provide a good corrective to such idealisation from a distance.

*Debates relating to the relationship between the individual and society*

Perhaps one of the most important debates in much of the social sciences concerns the relationship between us as individuals and the societies of which we are members. Integral to the debate is trying to understand whether we as individuals shape the way society is organised and patterned (its characteristics) or whether society's structures and systems (for example, its economic system, its system of government, its system of religion) shape us as individual members.

There is little doubt that we learn our values and so on by being members of a particular society, and this accounts for the tremendous diversity of social characteristics globally. But when we actually try to ascertain whether it is us shaping society or society shaping us, and how this occurs, matters become quite complicated. There has been much debate about this relationship, with perhaps a position now emerging that essentially sees the individual and society as so completely integrated that it is impossible, and not very helpful, to try to disentangle them.

On first appearances, this might seem to be stating the obvious. Of course we and society are integrated. But it has not always been as self-evident as some may think. Many people have different assumptions about the nature of the relationship between the individual and society. Some see the individual and individual behaviour being able to be dictated through the policy process, for example. Therefore, if you want to, say, conserve biodiversity through protected areas, a useful approach is to legislate. In effect, there is an assumption that you can legislate away, for example, local people's rights to part of the protected area. This suggests that an individual's (or group's, or culture's) behaviour and values are merely the result of society's influences. If you legislate to change a part of that society, you can alter behaviour. But does it? No.

In much the same way, individuals are not the sole creators of society's values and world views. An important area for global biodiversity conservation does not become protected merely because various people think it is a good idea, and it does not remain protected because some people think it is important. It does require something which is more than the sum total of people's good intentions and actions. If we change the example above to look at it from this perspective, there may well be a simplistic idea that protection of an area can be achieved by focusing only on local people. For example, it would not be too much of an overstatement to suggest that some people assume that through, for example, environmental education and changing individual behaviour, society can change. Others may view such an idea as, at best, naive.

Therefore, it is of utmost importance to recognise that conservation dilemmas, land and natural resource use and the ways in which we perceive and use the natural world are the result of a complex interaction between the individual and society. The ways in which projects have addressed this relationship, and its importance to integrating conservation with development, to protected area management and use, and to the broader goals of biodiversity conservation, are explored in more detail throughout this book.

*The state and the political process*

According to Ham and Hill, policy analysis describes a range of activities, its importance being in the 'doing' rather than the 'defining' of what it is (1985: 3). However, they make an important distinction between analysis *of* policy and analysis *for* policy. This distinction is 'important in drawing attention to policy analysis as an academic activity concerned primarily with advancing understanding and policy analysis as an applied activity concerned mainly with contributing to the solution of social problems' (1985: 4). In other words, there is a distinction between theory and its use in practice.

It is important to note that, whilst there is a distinction made between the two uses of policy analysis, together they form an integrated approach to the subject. That is, both dimensions contribute to our understanding of the policy process and its uses.

In the search for integrated conservation and development, protected area managers are operating within a policy context. Understanding the policy process is more than an academic exercise. It is central to the process and the delivery of policy which achieves its aims – in this case the aims of biodiversity conservation through protected area management, use and formation. An understanding of the policy process opens up a capacity to assess the relative power and influence of lobby groups, to lobby in your own right or with the support of other players, to advise local people on policy direction and implementation and to act, where necessary and appropriate, as their agent. The 'analysis *for* policy' dimension is therefore an important one.

Equally important, however, is the 'analysis *of* policy' dimension. Protected area managers have an increasing role in the assessment of policy initiatives. They are in a unique position to see the result 'on the ground', using evaluative frameworks which highlight the benefits of an integrated approach to protected area management. Their role, therefore, can be both proactive and evaluative.

### A final comment on sociology and anthropology

Protected area management, while obviously concerned with managing ecosystems, is fundamentally about the management of people, their aspirations and their relationship with nature. It is about understanding and assessing the relationship between society and nature, and about social conflict and social cohesion. From this perspective, it is very much about using the insights of sociology and anthropology, not at the end of a project or at the first signs of problems with people in the resource equation, but as a central and integral component of the establishment and continuation of management regimes and approaches.

Anthropology and sociology, as academic disciplines, have been quite tumultuous in recent years. Many anthropologists and sociologists have attempted to reassess the disciplines by reflecting on their past and present complicity in state relationships of power and domination of local or minority peoples.

Anthropologists today are usually fully aware of these issues and now practice a much more critical or reflexive anthropology. Anthropological study is now used as a form of 'cultural critique' (Marcus & Fisher 1986), or even a policy-informing discipline. Indeed, the practice of anthropology has changed in recent times; indigenous third and fourth world peoples have become empowered and now in many instances exercise the right of not allowing patronising or irrelevant ethnographic work in their communities. In many cases, indigenous and other communities set research agendas and hire their own anthropologists to pursue land rights claims, or to advise on development projects. Beyond this, some indigenous and other local people have studied social science themselves, including law, conduct their own 'insider' research and provide their own advocacy (Brettell 1993: 14).

Similarly, many anthropologists now attempt to form more collaborative relationships with those 'studied'. As communities demand that their cultural and intellectual property rights be honoured, anthropologists are having to negotiate what knowledge and information are taken from a community, what is published and how it is said (Greaves 1995).

Sociology in its turn has become more frequently applied with the recognition that sociological knowledge can and should be used to uncover power relations and the social basis of inequality. Many sociologists see their roles as outside change agents where there is no antiseptic distance between the sociologist and what she or he studies.

Increasingly also, some anthropologists and sociologists argue that those with power, such as multinational corporations and international aid agencies, should be studied rather than those who lack power. The notion of 'studying up' has become emphasised since the 1960s and informs at least some anthropological and sociological work (Wright 1994). Mark Hobart (1993) and his colleagues, for example, have critically examined the development process, its language, assumptions and practices, especially the part 'played by Western scientific knowledge' and how its application in some situations disempowers local people. Other anthropological work has framed the human dimensions of food policy in Africa and Latin America (McMillan 1991).

To summarise, anthropology and sociology have many applications in today's world and are well suited to studying the complex interrelationships of people and culture to the environment. They similarly have much to contribute to an understanding of local people, development and protected areas. The anthropological perspective provides a fine-grained understanding of management and development agency cultures, and how these cultures articulate or fail to articulate with local peoples and cultures. Similarly, the sociological perspective provides a much needed structural or systemic analysis of society and its components. Anthropological and sociological understandings form an essential part of coming to terms with any local level development projects.

## The Use of Economics

### The environment and economics

At first glance, some people would question any consideration of economics in issues where social and environmental values appear paramount. To understand why it is important to take into account economic considerations, we must look at the way the environment and human systems interact. Whilst this has been introduced in our discussions above from a sociological and anthropological perspective, this section examines similar issues from an economic perspective.

Tisdell (1991: 2) defines economics as the 'science which studies the allocation of scarce resources in society as a means to the satisfaction of human wants or desires'. Economics is also a framework which allows us to understand the interactions of resources in society, trade and market resources, and compare the values of different resources or different actions. Environmental resources, including environmental qualities, are part of this framework – no separation can be made between the environment and the economic system. Pearce *et al.* emphasise that *'The economy is not separate from the environment in which we live.* There is an interdependence both because the way we manage the economy impacts on the environment, and because environmental quality impacts on the performance of the economy' (1989: 4).

The environment, and the quality and abundance of environmental resources, including biodiversity, act on the economic system in a number of ways: as the overall biological framework which supports humanity; in the supply of resources for human use; in the absorption and degradation of wastes; and as an intrinsic good in their own right. On the other hand, economies act on the environment by using resources, by producing wastes that are (ideally) absorbed and modified by natural systems, and by changing aspects of the natural environment. Some people argue that rather than the environment fitting into the economic system, the economic system should be seen as fitting into the overall environment or ecology. Whichever point of view is accepted, economics is clearly an important consideration.

This interaction means that, in reality, any policies or actions made for economic reasons will have an environmental effect, and any made for environmental reasons will have an economic effect. When the interactions with other parts of social systems are taken into account (a political decision may affect an environmental protection policy, which may affect a local economic structure, which may disadvantage a local community, for example) the need to address issues from a holistic or at least a multidisciplinary perspective becomes clear. Protected areas and development issues are therefore inextricably linked to economic systems.

## *Economics and environmental problems*

The dominant school of economic thought today is welfare economics, which is concerned with maximising the welfare of society measured by summing the welfare of individuals within society. Welfare economics assumes that the most efficient way of allocating resources amongst individuals is through efficiently functioning free markets. While the market system is supposed to allocate resources optimally, it does not always succeed in doing so and this is called market failure.

There are a number of ways in which market failure can occur. Environmental goods, because they usually are public or partial goods, often cannot be allocated by the market system. Because no user has to pay for increased use of the resource, and because there is an incentive to use as much as possible (because others will do the same), individuals or groups acting in self-interest can overuse and degrade the good.

Another way in which the market system can fail is when the consequences of actions fall outside the market system. In some situations a person or group making a decision does not bear all of the consequences of that decision (this occurrence is called an externality, because the consequences of that decision are not included within the market system). For example, if a company makes a decision to harvest timber from a forest, they may not be the ones who suffer from the consequences of changing the natural resource. Local people may suffer from loss of hunting opportunities or other uses of the forest, and from deterioration in water supply from the forest catchment. People further afield, and society in general, might lose the benefit of knowing that the area exists in its natural state, and may have lost benefits from medicinal or scientific discoveries related to the forest. Whilst the assumptions and indeed the theoretical positioning of market economics are both the cause and the legitimation of these types of actions, those bearing the costs of the destruction of protected areas and habitats are often not the groups which reap the benefits.

Many environmentalists have been very vocal in their criticism of welfare economics. Alternative schools of thought, such as humanism, the naturalistic or ecocentric ethic and the variety of non-capitalist systems, all provide different outlooks and different means of dealing with environmental problems. Neoclassical welfare economics, however, is at present the dominant paradigm in which most decision making and policy development takes place.

For those involved in the search for integrated conservation and development alternatives, it is important to gain an understanding of these assumptions (and the methods associated with them). Not only will this provide a means by which the dominant paradigm can be further understood, it also provides ways to work within and reinterpret the paradigm as appropriate to local circumstances.

*Environmental economics*

Because of the problems with conventional neoclassical economics, and the obvious environmental deterioration that can occur as a result, a subdiscipline that addresses environmental issues from the perspective of economics has developed. Environmental economics contributes analytical tools and methods at many levels, from the global to the local. Within the subdiscipline, a great deal of attention has been focused on a number of important areas. Some of the areas that have been developed and researched at length include: methods of informing decision making and policy making for the management of environmental goods; non-market valuation methods (to allow goods and resources with no prices to be valued and therefore incorporated in economic analyses); instruments and regulations to allow environmental values to be included or better represented in the market system (including correcting externalities); and the theoretical analysis of the economics of environmental problems.

### Ecological economics

As many theorists and environmentalists feel that the restrictions of the neoclassical paradigm are fundamental to environmental problems, and that it is necessary to move wholly away from the paradigm, an ecological economic framework has been developing. Ecological economics investigates the limitations of the neoclassical framework for making decisions about ecological systems. Methods by which environmental decisions could be made are evolving, based on a systems approach that determines the capacity of ecological systems, and then uses this capacity as a constraint on the options for economic activity. Ecological economics is concerned with achieving sustainable outcomes with the minimum economic cost, while environmental economics, it is argued, is concerned with maintaining economic activity with minimum environmental cost.

### Economics, protected areas and development

Given that many environmental problems can be caused by flaws in economic systems or economic management, and that environmental policies may have adverse impacts on the economic wellbeing of local people (and the wider community), there are methods of balancing these issues, and tools to help managers, policy makers and local people do so. The economic context is important to the success of management policies, development projects and community change related to protected areas.

An understanding of the local, regional and wider economic as well as social context of development or change is necessary, as unforeseen effects of policies or induced change may manifest themselves through economic structures. There are usually potential economic benefits related to local development and protected areas, and it is fundamentally important to ascertain who should benefit from development; who will benefit from development; and who, if anybody, will suffer from development.

Development projects must be implemented so that people at the local level benefit. If this process occurs, as it should, by way of partnerships between local people and other actors, a reconceptualised notion of what development entails may emerge from the process. This may be the antithesis of neoclassical notions of what constitutes development.

## IMPORTANT SOCIAL SCIENCE CONCEPTS

As in most areas of study, the social sciences have developed a language of their own in order to provide a shorthand way of making sense of complex issues. To make the most of social science understanding, it is important to grasp some of these key terms and concepts which we now briefly summarise.

### Social Science Terms

*Agencies of socialisation*

These are social contexts in which socialisation takes place. The socialisation process plays an important role in an individual's use and perception of the environment. Agencies of socialisation, such as the family, formal or informal education, the government, the community and systems of religion, all can and do impact on this. Agencies of socialisation are therefore crucial to acknowledge in any process of local level development. Community education programmes, discussions with tribal elders, intervention in local political and religious systems and other such approaches have been used to facilitate an atmosphere conducive to integrating development and conservation through protected areas.

*Alienation*

This is the sense that control over our own abilities and our future is not within our grasp. The importance of alienation can be understood in two key ways. First, by recognising the social authorship of conservation/development issues we can understand our role in them, as well as our potential strengths and limitations. Secondly, local level development is precisely about overcoming the sense of alienation that local people often feel. Local people are empowered to participate more fully in the conservation/development approach, feel more ownership of the process as significant and important actors and, ultimately, provide the mechanism for the protection of biodiversity.

*Common-sense beliefs*

By this we mean widespread beliefs about the social and/or natural world held by a society's members. Whilst this seems obvious, common sense is not all that common. What constitutes common sense is, in fact, a result of the norms and values of communities, groups or societies. It therefore changes according to local customs, expectations and the like. The result is that what is common sense in one setting may not be common sense in another. If we act on the assumption that common sense, as we see it, is universal, then we run the very high risk of forcing our perceptions and values on those who don't share them. Therefore, we start to develop inappropriate projects, see only what we expect to see or, worse still, believe that our views (our 'common sense') are better or more correct than those expressed by others.

## Conflict

Conflict can be defined as antagonism caused by a clash of cultural, social, economic and/or political interests between individuals or groups. Integrating development with conservation through protected areas can be an act of conflict resolution as various key actors may have a broad range of interests which they may want to protect.

## Culture

When social scientists use the term 'culture', they describe a vital concept of their discipline. For them culture means the entire way of life of a society. 'Culture is all the shared products of a given society: its values, knowledge, norms and material goods' (Furze & Stafford 1994: 29). Every individual who is an active member of a society must share in its culture, although sharing is never complete or fully predictable, and may vary along lines of power and inequality, gender and ethnicity. Thus, cultural knowledge is unevenly shared and often contested. Culture contains certain elements, including norms, values and ideals, and can be divided into sub- and countercultures. Culture includes the non-material (knowledge, values, beliefs and social norms) and material (arts, crafts, clothing, dwellings, tools etc.) aspects of society.

## Economics

Economics is the '(social) science which studies the allocation of scarce resources in society as a means to the satisfaction of human wants or desires' (Tisdell 1991: 2). Economics is also a framework which allows us to: understand the interactions between labour, natural resources and capital; trade and market resources; compare the values of different resources or even the economic values of different actions.

## Economic efficiency

To achieve economic efficiency is to have allocated resources within society in the optimum way – in such a way that society obtains the greatest benefit for the least cost in this allocation.

## Economics – welfare

This is a school of economic thought (dominant today) that sees the role of economics as maximising the welfare of society through improving the welfare of individuals within society.

## Economics – ecological

Some economists and environmentalists are critical of the neoclassical economic framework's inability to deal with social and economic issues adequately and equitably. Work has been conducted grappling with these shortcomings, and attempts are being made to develop alternative frameworks. The emerging ecological economic framework is systems based, with the capacity of ecological systems determining the level of economic activity acceptable, and the economic merits of alternative options analysed under environmental constraints.

## Economics – environmental

Environmental economics is a subdiscipline of welfare economics. It addresses environmental concerns from the perspective of economics. An environmental economic framework facilitates the incorporation of environmental economic concerns into conventional economic analysis, thereby improving decision making at the economywide, sectoral and micro levels.

## Economic value

Value, as used in economics, is broadly defined as anything which adds to human wellbeing. The benefits obtained from the consumption of commodities (such as minerals and timbers) add to an individual's wellbeing, as do the unpriced benefits received from environmental amenities (such as national parks and clean air). In this sense, anything from which an individual gains satisfaction is deemed to be of value so long as the individual is willing to give up scarce resources for it. If we define a 'good' as anything that has economic value, then this value can be thought of as the amount of money a person is willing to pay to obtain the good, or the amount of money the person requires as compensation for the loss of the good.

## Ethnocentrism

Ethnocentrism refers to the tendency to look at other cultures using the norms and values of our own as a frame of reference. If this happens, social scientists may view their value positions, ideas and lifestyle as being more important (or better) than that of another group or individual. This sets up an 'us' and 'them' or 'insider/outsider' mentality and when this occurs, meaningful participation in the local development process is unlikely. To achieve social science understanding, it is important to remember that norms, values, ideas and ways of living are products of social, cultural and political complexities.

*Globalisation*

Globalisation refers to the spread of social, economic and political relations worldwide. We have seen this happen as a result of the modernisation/industrialisation process of development, but we also see it in other contexts such as the globalisation of media and mass communications with the resultant global marketing. The importance of globalisation as a concept stems from an acknowledgement that local level development will have to balance local with national, regional and global interests and forces.

*Goods*

Goods are products or elements which provide humans with satisfaction. This satisfaction may be, for example, economic, aesthetic, cultural or use related. In the context of economic analysis, however, all of these components are regarded as having an economic value.

*Goods – environmental*

Goods that exist not as a result of human production (directly or indirectly) but that occur naturally are often referred to as environmental goods. These may include landscapes, habitats or natural systems, and components of such systems, including species and populations. Raw materials that are part of, or produced by, such systems as air, water or soil are also environmental goods. The satisfaction which people derive from environmental goods can be varied, including productive or consumptive use of resources which gives satisfaction, pleasure in using a natural area for recreation, extractive uses and scientific or medicinal uses (all of which are examples of use values). In some cases, deriving satisfaction may not involve uses at all – for example, gaining satisfaction from the knowledge that a natural area exists (existence value), or from knowing that the area will be available in the future to use (option value) or to leave for others (bequest value). Most environmental goods are public or partial goods.

*Goods – public and partial*

Public goods are goods that cannot be restricted in provision to any individual or group within society – they are available for all, and the use of the good does not reduce the amount of the good available for others. Partial goods are those where, although the good is available for all to use, use of the good may change the quality available to other users after a certain point. For example, a national park is available for members of the public to use, but increasing use may, through congestion or degradation, reduce the satisfaction available to other users. Importantly, such public or partial goods are very difficult to trade in a free market and hence it is difficult to estimate their true economic value.

## Ideology

Ideology is the shared beliefs, values or ideas which serve to justify the interests of groups. It is an important concept. The legitimation of particular groups, or even models of development, is a result of ideology. The belief in the usefulness and responsibility of integrating local people into the conservation/development agenda is premised on a number of factors and the legitimation of this approach comes from the ethical positions discussed in Chapter 1. Those of us who see local conservation and development as desirable therefore share an ideology governing this. Likewise, those who may not see it as desirable (because, for example, it may threaten their social, economic or political interests) will also have an ideological justification for their stance. It will, of course, be different to ours. In a sense, then, we debate our position, others debate their position, and what results is the spread of particular beliefs to other groups as they are convinced or swayed by the force or logic of groups with competing beliefs.

## Legitimacy

Legitimacy is the belief in the validity of the existing social and political order. There are obviously many ways of legitimating a social or political order. The important point is that legitimacy is very powerful, and that it governs a wide range of beliefs and actions. The integrated conservation and development approach, if legitimated at the local level, has much more chance of being successful than if it is not viewed as a legitimate form of intervention.

## Local knowledge

Local community knowledge is possessed by people – knowledge about ecosystem processes, community organisation and structures. It is at the heart of the local level development approach.

## Market system

The market system is driven by individuals acting in self-interest, creating demand for goods or services which will help them obtain satisfaction. The market system is a set of institutional and cultural arrangements which guides the allocation of resources by assigning monetary values. One reason for the market system being dominant in economics is its capacity (under certain conditions) to allocate resources in an economically efficient way.

## Paretian efficiency and potential Paretian improvement

Determining economic efficiency by measuring individual benefits objectively and comparably between individuals is difficult, and so an alternative way of defining economic efficiency is often used. Rather than finding out how one action will affect the individual utility of all people affected, the criteria is adopted whereby any action that makes people better off while making nobody worse off is considered an improvement (Paretian efficiency, after Vilfredo Pareto, 1848–1923). The ideal state where it is impossible to make any one person better off without making someone else worse off is called Pareto optimality. In reality, many situations exist where changes make some individuals worse off and others better off. The concept of potential Paretian improvement (or the Kaldor-Hicks criterion) was developed in which, if those who gain from a change in resource use could compensate those who lose and still remain better off, the change is considered an improvement. These criteria provide a goal for decision making – if the potential Paretian improvement criteria were to be used in all cases, society could, on the whole, only benefit. The questions of equity and distribution of benefits are not, however, addressed by these criteria.

*Power*

Power can be defined as the ability of individuals or groups to further their own interests. Once again, this is a key concept in social science and also to the integrated conservation/development approach. Societies and communities are not homogenous entities. They are hierarchical, where individuals and groups have influence over others. This influence can be legitimate or illegitimate, traditional or a result of non-traditional forces. It does, however, exist. It also can (and does) impact on how a project will work, if it will be supported and, if so, by whom.

*Self-interested rationality*

The theory of self-interested rationality dictates that individual actors in an economic system (such as individuals, firms or organisations) will act to their own advantage (maximising their own utility).

*Social change*

Social change is represented by the alterations to the structures of society over time. Often, what the local level development approach is doing is trying to intervene in, and thereby govern or influence, the processes of social change occurring locally, nationally, regionally and globally.

*Social institutions*

These express the organisation of social, economic and political activity which is followed by the majority of members of a society. They most often include the family, the education system, a system of religion, a political and economic system and the more general patterns of community norms and values.

## Social stratification

Social stratification is the structured inequality of groups of people based on their access to valued social, economic and political resources.

## Social transformation

This is the process of social change which occurs within societies or groups within societies.

## Socialisation

Socialisation is the ways in which individuals learn the values and norms of a society.

## Society

A society is 'a group of people who live in a specific territory and are subject to a common political authority. They are aware of being different to those societies and cultures around them' (Giddens 1989: 731).

## Sociological imagination

This is the process by which individuals 'think themselves away' from their familiar social understanding (Giddens 1989: 750).

## Values

Values are the ideas individuals have about what is good and bad, right and wrong, important and unimportant, beautiful and ugly, harmful and safe, just and unjust.

# 4 Using Methods from the Social Sciences

The social sciences offer a variety of methods for use in the development of social understanding. An important dimension to the use of these methods, however, relates to the complexity of social processes which need to be understood. It is impossible to reduce the complexity of human affairs to iron laws of cause and effect. What is possible, we would argue, is an approach whereby social understanding can be achieved at a level of 'optimal ignorance'.

The idea of optimal ignorance highlights the assertion that deriving knowledge and understanding of social, economic and political processes is at best an imprecise endeavour. This is not to say, however, that meaningful understanding is not possible. Rather, it implies that our understanding has limitations and we should bear these in mind when using social science knowledge. Social science research has direct impacts on the ways people live their lives and can be intrusive. It is therefore vital that it is done well.

In this chapter we present overviews of some important methodologies and techniques in social research. First we describe methods and techniques from the social sciences. We then go on to look at methods of economic analysis and conclude by exploring approaches to programme evaluation. As you read through the chapter and reflect on what it is saying, it is probably worthwhile considering the methods and techniques in the light of the social science enterprise, especially in relation to the importance of combining these with your broader social understanding.

The process of social research involves an interaction between researchers and people. The form and structure that this interaction takes vary with different research methodologies; some are highly structured and controlled by the researcher, while others are less structured and in a sense controlled by those the researcher wishes to learn from. This variety highlights the spectrum of social science methodological approaches and their differing philosophical and ethical foundations.

Please note that what follows is not a 'how to do' social science research. Rather, it aims to introduce you to a variety of methods which social scien-

# USING METHODS FROM THE SOCIAL SCIENCES

tists use, as well as their strengths and limitations. It is aimed at familiarisation and not at equipping you with a comprehensive and sophisticated knowledge of each one.

There are a variety of terms used in social science to refer to the people involved in the research process. The person collecting information or conducting the research is usually referred to as the researcher. The use of this term goes beyond the strict research definition. Researchers are not only 'outsiders' who study 'others'. They are part of the research process and the quest for social understanding. Therefore, researchers can also be local people, project managers, protected area managers, as well as the 'outsiders' (see, for example, Chapter 6).

The people from whom information is being collected have a number of different titles – for example, respondents, informants, subjects or partners. The different titles are generally assigned according to the methodology being used, and reflect the role that the researcher feels the other is playing. 'Respondent' is used when referring to the people to whom surveys or questionnaires are sent or 'applied' – the title denotes the role that they play in the interaction, responding to preordained questions. 'Informant' is used to refer to people who are interviewed in ethnography or in-depth interviews – this title reflects the desired process of 'teaching the researcher'. 'Subject' is used to refer to those who are the focus of hypothesis-testing research – where a researcher has a theory that is confirmed or denied by testing the responses of a person; again, the title illustrates the role of the person, 'subjected' to a process. 'Partner' denotes a much more interactive and action-oriented approach to social understanding. The relationship between, for example, a researcher and local people operates on a cooperative basis, usually framed by problems identified through this cooperation.

A distinction is also made between techniques and methodologies. A research technique (for the purposes of this book) includes activities that are used to collect or analyse information. A methodology, on the other hand, may be a collection of techniques brought together in a certain way that can be used to investigate an issue from a particular perspective. A variety of methodologies may be used for the same issue. These may include the same or similar techniques. However, because a different interpretive perspective is used, the outcomes of the research may be different.

## QUALITATIVE AND QUANTITATIVE RESEARCH

There are two forms of research fundamental to the social sciences – qualitative and quantitative. Minichiello *et al.* (1990) explain the distinction between qualitative and quantitative forms of knowledge in two ways – the 'conceptual' and the 'methods used'. In the conceptual sphere, qualitative research is 'concerned with understanding human behaviour' from the perspective of those particular people involved. Qualitative research also 'assumes dynamic and negotiated reality'. In contrast, quantitative research is conceptually 'concerned with discovering facts about social phenomena' and 'assumes a fixed and measurable reality' (Minichiello *et al.* 1990: 5).

What does this mean? In essence, some researchers (the interpretivists) believe that it is important to recognise that people interpret their own lived realities according to their social, economic and political contexts – lives, experiences, perceptions and surroundings. Other researchers (the positivists) focus on 'facts' as observable, measurable entities that can be assessed objectively and that have meaning beyond their immediate social context. That is, they look for social laws which govern human behaviour and interaction.

Research methods also reflect this conceptual distinction. Qualitative information is collected through methods such as unstructured and semi-structured (in-depth) interviewing and participant observation, and is 'analysed by themes from descriptions by informants'. Quantitative information, on the other hand, is 'collected through measuring things' (using survey instruments, observation, structured interviews) and is 'analysed through numerical comparisons and statistical inferences' and is reported in the same way, using statistics (Minichiello *et al.* 1990: 5).

### The Choice between Research Methods

On the surface there appear to be many different techniques and methodologies for social research. Byers, citing White, attributes this impression to the 'disciplinary and methodological dogmatism associated with many social research methods' (1994: 21). Essentially, 'all field techniques are based on a combination of three complementary approaches: observing, listening and asking questions, which cover a spectrum from researcher structured to respondent and/or situation structured' (Byers 1994: 21). This spectrum, according to White, also represents a trade-off between the quantitative and qualitative approaches (Byers 1994).

While it is recognised that the values of a researcher will affect the approach used and the interpretation of its results, it should still be possible to identify situations which require either the more intricate social understanding that is achieved through qualitative research, or the more generalised, numerically based understanding that quantitative research can bring. It may be that a sponsor or funding agency requires a quantitative understanding of a situation, while local people require a qualitative understanding of a situation in their own social context. The use of different methods where possible to provide new information and cross-checks or triangulation of information from different sources is also advocated by White (Byers 1994) – a technique adopted in the participatory and rapid rural appraisal process.

## SOCIAL RESEARCH METHODOLOGIES AND TECHNIQUES

The following methodologies and techniques exemplify different approaches to social research. The methodologies discussed – ethnography, participatory and rapid rural appraisal and social impact assessment – are used by social scientists to achieve a broad, integrated understanding of a culture, issue or problem and so on. We describe the strengths and weaknesses of each methodology to highlight their suitability for various social research activities. We then look at three of the main social science techniques – surveys and questionnaires, interviews and focus groups – in greater detail, as these techniques are used in a variety of methodologies. In a section such as this we could not hope to provide a detailed discussion of the complexities of all social science techniques. However, the information presented here should provide readers with enough understanding of terminology and technicalities to be able to understand and critically review research reports and projects and commission new research that may be of interest.

### Methodology 1: Ethnography

At one level, ethnography can be simply described as the work of describing a culture with a central aim of developing cultural understanding from the perspective of those who live it (Spradley 1986). One of ethnography's major assumptions is that of cultural relativism. This concept emphasises that we cannot understand another culture using the values of our own. If we impose our values on what another culture does, we are being what is called ethnocentric, which means believing that our culture is superior to another. Ethnocentrism, and the similar term racism, have been used by many powerful groups to legitimate such concepts as domination, colonialism etc.

An ethnographic approach means that the researcher becomes a student, and the researcher comes to the field with what Spradley has described as an attitude of almost complete ignorance. As a result, the type of knowledge you develop using the insider's view is different to that which you would have if you were to use your own understandings to interpret cultural situations (Spradley 1986).

There are two important dimensions to the ethnographic approach which we will emphasise here. The first relates to ethnography as methodology. This is concerned with 'doing' ethnographic research. The second relates to ethnography as politics. This addresses the ethical and power-based dimensions to ethnographic research, and provides a context within which the method of ethnography can and does occur. These two dimensions are important, but the political context is perhaps more so, given that the aim of a participatory approach to an integrated conservation and development project is an ethical position and can be an attempt to reconstitute power relations.

### Ethnography as method

As mentioned above, ethnography is a process of describing culture from the perspective of those who live it (the 'insiders'). Describing culture or social reality from an 'insider's' perspective requires specific data-gathering techniques. Because of the nature of the ethnographic approach, data gathering techniques need to be flexible in order to accommodate the diversity of possible situations likely to be encountered. Later in this chapter we discuss some of the possible techniques (see, for example, interviewing and the participatory or rapid rural appraisal approaches). Interviews are generally used in conjunction with participant observation or other ethnographic research techniques. The techniques are used to obtain behavioural and attitudinal data covering a range of cultural aspects, which will allow the researcher to have a greater understanding of how the groups function, the world view of their members and, importantly, how they are reacting or may react to change.

*Participant observation* involves becoming an 'insider' by participating in the 'insider's' culture. Through this participation, observations are made of what is going on and notes (often in the form of logs or diaries) are made of these observations. Of course, with this approach, gaining legitimacy within the community and rapport with the insiders is essential to the success of the project. Gaining entry to the field can be a time-consuming process, as the researcher may be met with fear, suspicion or curiosity. In order to carry out successful participant observation the researcher must become accepted by the community to a degree where her/his presence does not affect the activities of community members. Issues related to gaining entry to the field are expanded on in Chapter 5.

There are many advantages of participant observation as an ethnographic technique. One of the most significant of these is the ability of the observer to witness the actual behaviour of the members of the group. Other methods rely on the accounts of the individuals themselves, which may not be accurate, as they might be altered to comply with socially correct practices or to satisfy the researcher's requirements as the informant sees them. The participant observation approach also allows the researchers to gain a sound understanding of the culture in which they are living, as they are immersed in the culture and witness its impact on the daily lives of the community (Ferraro *et al.* 1992).

While participant observation provides the researcher with insights which are impossible to obtain from non-participatory methods, there are a number of problems associated with the method. These include the intrusiveness of the researcher – the impact he or she has on the individuals being studied. The presence of the observer may change the activities of the individuals and thereby result in a misunderstanding of the group's cultural characteristics. Also, the method only relates to a very small sample size, and the results are difficult to generalise to other groups. As with all ethnographic research methods, the objectivity and bias of the researcher are issues which must be considered.

*Ethnographic mapping* is often used in conjunction with participant observation. It allows the researcher to learn about a culture by determining how individuals relate to their physical environment. Ethnographic mapping can reveal a variety of cultural characteristics, for example the ways in which people rely on various physical features such as forests, rivers and mountains, and how communities divide and distribute land (Ferraro *et al.* 1992). Importantly, it can help assess the strength of connection between people and their land and the resources which grow or live on them.

*Rapid ethnographic techniques* are becoming more commonly used to provide quick social understandings of groups involved in development projects for conservation activities. For example, the United States Parks Service has developed rapid ethnographic assessment procedures (REAP) to assist social impact assessments within protected areas (US Parks Service 1993). REAP recommends a combination of ethnographic methods including focus groups, transect walks and community mapping. REAP is designed to be implemented within four months, and assists protected area managers in satisfying legislative requirements to determine the social impacts of alternative management options. However, this rapid ethnographic study is designed for specific purposes and is not a substitute for prolonged studies involving other ethnographic methods.

### Ethnographic Fieldwork – Mapimi Biosphere Reserve, Mexico*

Andrea Kaus spent seventeen months in 1988–89 and two months in 1990 engaged in ethnographic fieldwork at Mapimi Biosphere Reserve in Mexico. She utilised a range of ethnographic methods and techniques during her study, including participant observation, interviews, surveys and ethnographic mapping. Her work was implemented as a hypothesis-testing study, and resulted in a number of suggestions of future management options for the reserve. The fieldwork was designed to collect baseline data and to engage in participant observation on the reserve. Two distinct groups emerged within the reserve's population – researchers and residents. Kaus spent time living with each of these groups, and gathered data which allowed her to compare the different world views of each group. The participant observation component of the ethnographic fieldwork was

---

* Andrea Kaus carried out fieldwork in the Mapimi Biosphere Reserve from 1988, and produced a thesis on characteristics of the reserve entitled *Common ground: ranchers and researchers in the Mapimi Biosphere Reserve* (Kaus 1992). This case study is based on her thesis. See Chapter 10 for further information.

also used to determine both residents' and researchers' work routines, patterns of land use and resource management techniques.

A property survey and a census were carried out on all of the residents of the core and buffer zones of the reserve, and on a portion of those from the transition zone. The survey and census were implemented through interviews, which were mainly informal. The property survey derived information regarding: 'the type of property, the amount of land, the number of residents and landholders, their principal place of residence, the date the property was bought or established, property boundaries, administration and management, cattle brands, type of economic production, water resources, infrastructure (buildings, roads, electricity), and access to social services (schools and medical help)' (Kaus 1992: 81). The census covered information such as: 'names, age, sex, dates of birth, length of residence in the settlement, previous residence, education, occupation, and the residence of their parents' (Kaus 1992: 82). The survey and census were carried out at household level and the survey information was gathered from a male member of the household, while the census information was gained from a female member. This division was suggested by the residents, and revealed a gender-based distribution of responsibility and knowledge.

Five residents of the reserve were asked to maintain diaries, in which they recorded their daily work activities. These diaries were kept from 1989 to 1991 to give further insight into the working habits of the residents. Ethnobotanical knowledge was not easily gained from any of the above methods, so ethnographic mapping was used to provide an insight into the local perceptions of the environment and of the conservation programme being implemented within the reserve. The mapping exercises revealed local names and perceptions of land division, location of plant resources, range management techniques, elements of local history, and opinions regarding the reserve.

Through utilising a range of ethnographic research methods and techniques, Kaus obtained a comprehensive understanding of the social and economic structure of the population of the Mapimi Biosphere Reserve. The research has led to the establishment of similar research programmes in other reserves, to allow for comparison of land tenures, economic activities and social relations.

*Ethnography as politics*

Historically, ethnography has been seen as a way in which outsiders gain social understanding of different cultures. Often these cultures have been described in terms such as strange or exotic, where social scientists (often western or western trained) study 'their' people and write up 'their' knowledge and understandings in academic journals and/or books, often in a romanticised way. It should be clear that this approach to social sciences has the stench of westerncentrism (and, at worst, neocolonialism) about it, and perpetuates the depiction of 'the other' exotic world which was the foundation of much of the social understanding of the colonial period.

Whilst much recent ethnographic social science has explicitly attacked these assumptions and this approach, a fundamental issue is still important. How do we reconcile the cultural meaning systems of the observer with those of the observed? What may occur with ethnographic accounts, to provide a simple example, is that the observer (in many cases, the 'outsider') has certain preconceived ideas about what she or he is going to look at (and look for). For example, we may go into the field to observe certain behaviour related to natural resource use. We may see some of the resources used by local people in terms of biodiversity conservation, or in terms of ecosystem management. The local people, on the other hand, may see the same resource but in different cultural terms. They may see it in terms of medicinal value, religious significance or of economic trade. We both see the same resource differently because we see it through our own cultural lenses.

The issue of cultural meanings may not be that (relatively) clear cut. The values, biases and cultural symbols of the observer can and do get in the way of their observations. As a result, it is difficult (if not impossible) to suggest that the observer is seeing the same things and interpreting them in the same way as those being observed. Life becomes very complicated when we assume that we do, and conservation projects are developed on the assumption that local people are acting in accordance with the interpretations of the observations of an outsider of (often) a different culture.

Too often, ethnographic accounts have claimed 'transparency of representation and immediacy of experience [where] writing [is] reduced to method: keeping good field notes, making accurate maps, "writing up" results' (Clifford 1986: 2). Ethnographic information should not claim to represent experience in this sense, because it only represents one view of the experience and that view is almost always that of an outsider.

Therefore, it needs to be highlighted that ethnography and observation are 'situated *between* powerful systems of meaning' (Clifford 1986: 2), the meanings that outsiders ascribe to their observations and those that insiders ascribe to their everyday reality. Ethnography can be a powerful social science method, but it can also lead to a distorted analysis of the ways people live their lives. Taking this to its logical conclusion, it can result in a project which fails to account for the needs of local people and/or reinforce the perceived distance between the management of a protected area and its relevance to local populations.

**Methodology 2: Rapid and Participatory Rural Appraisal**

Partly in an attempt to overcome the time and cost factors associated with long-term research into rural social, economic and political processes, rapid rural appraisal (RRA) methodology was developed. At the same time the approach to rural development was undergoing a transformation amongst some practitioners and organisations. The new approach sought to put people first or, in Chambers' (1986) telling phrase, put 'the last first'. The result of this change has been increasing participation of local people in the development process. The integration of this new ethical approach to rural development with the techniques of RRA has seen the emergence of new participatory rapid techniques widely referred to as participatory rural appraisal, or PRA. In PRA, social research methods are chosen which maximise the development of key social knowledge, minimise the gaining of superfluous information, optimise the cost-effectiveness of rural social research and ensure the participation of local people.

To achieve these efficiencies in the quest for social understanding, PRA uses a number of principles. It is these which provide it with its explanatory power and distinguish it from what Robert Chambers (1991) describes as 'quick and dirty' rural development research, or understanding which is derived in a short period of time using flawed or restrictive methods.

Essentially, the two principles which govern PRA are 'optimal ignorance' and 'triangulation' (McCracken *et al.* 1988: 12). Optimal ignorance suggests that it is not possible to 'know' the object of the research completely. Rural cultural, social, economic and political processes are so complex that the best researchers can hope to achieve is a type of informed ignorance about the worlds of their study. Because this is the case, it is not sensible, either economically or methodologically, to spend vast amounts of time in the field doing research which will have little additional value in comparison to time spent. Therefore, PRA techniques develop a variety of methods to achieve understanding within given time and financial constraints (imposed sometimes by funding agencies, by the reality of development interventions, by the 'objects' of study and sometimes by the researchers themselves). As long as the second principle, that of triangulation, is adhered to, optimal ignorance should be able to be achieved.

Triangulation involves the use of several different sources of information and PRA techniques to achieve social understanding. Therefore, this social understanding, framed by the principle of optimal ignorance, builds up through the use of diverse sources (McCracken *et al.* 1988: 12).

We may add here another important principle in PRA, that of the legitimacy of the local level. PRA emphasises the legitimate knowledge of local people as well as their integral role in the development process itself. It seeks to gain information from a variety of sources which include (and sometimes only include) knowledge derived from local people themselves. In other words, PRA is a two-way process, with learning and communication occurring between the researcher and the local level, in both directions.

But what type of understanding can PRA be used to achieve? When is it likely to be useful? According to McCracken *et al.* (1988: 9–10) PRA can be used to:

- assess the rural and other development needs of a community
- identify priority areas for further research and development intervention
- assess the social and technical feasibility of planned interventions in the rural development process
- implement and monitor development interventions

It should be noted that this list is not exhaustive. The uses of PRA are circumscribed only by the needs and imagination of its users, as long as its principles (which provide it with its explanatory power) are adhered to.

*Techniques of rapid and participatory rural appraisal*
There are many techniques which can be used in PRA. McCracken *et al.* (1988: 18–49) look at many: secondary data review; direct observation; semistructured interviews with key informants and/or with community groups; stories and portraits told by informants; uses of diagrams and transects to, for example, record changing land use patterns or changing environmental impacts; and workshops.

- *Secondary data review.* A review of published and unpublished data, which can take many forms, including surveys, studies, annual reports, trip reports, travel books, ethnographic literature, articles, maps, aerial photographs, satellite imagery, and computer data files.
- *Direct observation.* Personal visits and observations with time to follow up on what is seen. An observational checklist is one aid to systematic observation.
- *Do-it-yourself.* Much briefer participant observation than in the normal social anthropological mode can take the form of undertaking a rural activity oneself. This allows insights and prompts the volunteering of information that would otherwise not be accessible.
- *Key indicators.* Key indicators can be shortcuts to insights about rural social conditions and change, especially when suggested by rural people themselves.

- *Semi-structured interviews.* Informal interviews with checklists but without questionnaires, which permit probing and following up on the unexpected, without the requirement that all the checklist points must be covered in any one interview.
- *Key informants.* Identifying those best able to inform on particular topics, or to give special points of view, whether individually or as groups.
- *Group interviews.* Interviews and discussions with groups, whether casually encountered (such as coffeeshop or teashop groups); specialised or focus groups of similar people; structured groups with an organised composition to represent different points of view, capabilities, or knowledge; and community groups.
- *Chain of interviews.* Sequences of interviews, whether to cover knowledge of stages of a process (such as following a crop from land preparation through cultivation, harvest, marketing, processing, storage, sale, and cooking to consumption), or to follow through on similar topics from early to later contacts, as when group interviews lead to the identification of individuals who are key informants. Repeat interviews in different contexts, including walks with observation, can be part of this.
- *Transects and group walks.* Systematic walks (or in large areas, rides or drives), for example from the highest to the lowest point, visiting and observing diverse conditions en route, including the poorer people and microenvironments.
- *Mapping and aerial photographs.* The use of formal maps, whether general or specialised, and the preparation of informal maps based on observation and on local knowledge. The use of aerial photographs as aids to ecological, social, and political mapping and to identify longitudinal change.
- *Diagrams.* The use of diagrams to express, share, and check information. These include diagrams to represent information that is spatial, for example for transects; temporal, including trend lines for changes over the years, and seasonal calendars for changes within years for dimensions such as labor, diet, disease, cropping practices, prices, livestock fodder, rainfall, migration, and tree use; social, including links and overlaps between groups and institutions at the community level; and concerning processes. Types of diagrams include sketches, bar diagrams, histograms, flow diagrams, Venn diagrams, and decision trees. Venn diagrams (also known in South Asia as *chapati* diagrams) have been used to identify village institutions and their interrelationships.
- *Ranking, stratifying, and quantification.* Methods for eliciting knowledge and preferences from informants, which are both quick and enjoyable, have been used. Aids to quantification and ranking include

the Atte board, informal pie diagrams, and various systems of questioning. Wealth ranking has proved a quick and accurate method for stratifying a rural population and has been tested and found to be effective in several different environments.
- *Ethnohistories*. These are histories recalled and recounted by rural people. In one notable example, cultivator biographies were elicited with respect to a particular crop, cassava, in the Dominican Republic. In another, information was obtained on changes in child-rearing practices in Ghana, with three generations of mothers as informants.
- *Stories, portraits, and case studies*. These are anecdotes and descriptions of people and households, farming systems, social groups, villages, events, customs, practices, or other aspects of rural life, designed to portray conditions as one part or stage in understanding.
- *Team interactions*. The deliberate organization of team interactions is a part of many RRAs. The classic example is Hildebrand's *sondeo* technique, originating in Guatemala, in which social and biological scientists are paired, changing partners each day. This technique has also been adopted in Australia.
- *Key probes*. Sometimes an exceptionally revealing key question can be identified . . .
- *Questionnaires*. Late and light. If a questionnaire is needed, it is usually devised late in the investigation, tied in with dummy tables that are known to be needed, kept short and simple, and immediately analyzed.

Source: Chambers 1991: 523–6

Examples of PRA and RRA are given in case studies in later chapters. The Joint Forest Management programme in India used PRA methods widely, which are described in Chapter 9 (Poffenberger *et al.* 1992a,b). In Chapter 10 we describe an RRA undertaken by the authors in Xilingol Biosphere Reserve in China.

PRA is an innovative technique. It has to be because it deals with complex social, economic and political processes which require flexibility in project design and methodology. Precisely because of this, the above do not represent the final word in PRA techniques. Nor are they necessarily self-explanatory. For example, semistructured interviews may sound appealing, but how do you go about asking questions? Who can you ask them of? Are these people different to who you should ask them of? What happens when the same question asked of different people obtains a different response?

These important points serve to remind us of the difficulty in gaining social science knowledge. They should also, however, highlight the importance of gaining that knowledge. After all, it represents ways in which development and conservation can be more fully integrated with protected area management.

**Methodology 3: Social and Cultural Impact Assessments**

Social and cultural impact assessment methodologies offer a framework for assessing and predicting the probable effects of a policy change or development (such as the construction of infrastructure like a dam). Social and cultural impact assessments attempt to combine the use of objective (technical) information about a project with subjective (political, social interaction and so on) information, in order to understand the opportunities and constraints presented by local communities to a particular project. Social and cultural impact analysis has been used in a wide range of contexts, including industrialised and agrarian communities, indigenous and rural communities, although much of the readily available literature deals with the western, industrialised context. As social and cultural impact assessments are usually project based, they are often described as a defined series of steps, followed through in a process, in order to develop findings or outcomes which can then be used to modify or halt a project. One definition of social impact assessment is:

> Social impact assessment is an attempt to predict the future effects of policy decisions (including the initiation of specific projects) upon people, their physical and psychological health, well-being and welfare, their traditions, lifestyles, institutions, and interpersonal relationships. (D'Amore cited in Craig 1990)

There are no prescribed rules or methods for social and cultural impact assessment – an emphasis is placed on the fact that each project has a unique set of characteristics, and therefore the methods used in each case must be tailored for that particular assessment (Finsterbusch *et al.* 1983). However, a generalised description of the steps in the process would include assembling background data, checking the accuracy of this using local knowledge, using field research techniques for identifying probable undesirable social effects, measuring and understanding impacts, examining alternative plans and impacts, and recommending strategies for mitigating impacts (Interorganisational Committee on Guidelines and Principles for Social Impact Assessment (ICGPSIA) 1994). In addition to this main frame for assessment, the different stages of any project's 'life cycle' need to be understood, as different stages may generate different effects. A number of authors also address the fact that social and cultural impact assessments are at a 'point in time', and may be subject to change in attitudes or perceptions – in essence, that they may not be able to account for every impact or effect arising from a project.

The ICGPSIA of the United States Department of Commerce produced a set of guidelines for social impact assessment in 1994 consisting of a structured series of steps. The content of these steps will be discussed very briefly, to provide an overview of social impact assessment in a generalised sense. Each development project might use some or all of the steps to a greater or lesser extent and in varying sequences. It should also be pointed out that this particular approach was designed to be used for development projects in the USA. Undertaking social impact assessments elsewhere will need to use a different approach to accommodate varying policy settings, legislative requirements and cultural and political norms.

1. *The description and identification of the project or policy change* is straightforward. This step serves to ensure that the geographical area and the people likely to be affected are identified. The time over which the project is expected to run should also be identified at this stage.
2. *The development of a public involvement plan*, as well as allowing the public to be part of the project, can allow the project to be informed by the more thorough knowledge that the local people may have of the area and the potential impacts of the project. Methods advocated for the involvement of the public include interviews, discussions, surveys, formal and informal public meetings, and calls for submissions.
3. *Identification of alternatives* can occur after concerns of the public have been assessed. The nature of the project and the views of those who manage and control various elements of the project planning and development will affect the potential for reasonable alternatives.
4. *Baseline conditions* of the human environment should be assessed. This will involve the present, and historic conditions, and will include relationships of people with the biophysical environment, the historical context of the project, political conditions, attitudes, and population characteristics.
5. *The identification and assessment of the probable impacts* (as perceived by the affected people and the project managers) is the core of the assessment exercise. Public participation methods (as described in step 2 of the process) will help identify the range of social impacts. It is important to recognise who in the community is expressing views about impacts or effects, as bias and power structures within or surrounding a community may have an influence. In addition to the use of these methods, the review of social science literature may also be important. The assessment of the impacts identified is also carried out at this stage. Categorisation, the development of indicators of impacts, and measurements using these indicators are suggested. A number of criteria are also suggested for identifying which impacts are likely to be

significant, including: probability of the issue arising; number of people affected; span of impacts (over time and space); magnitude of the cost or benefits for those affected; likelihood of further impacts; capacity to mitigate the impact; and presence or absence of controversy over the issue.

6. *Projecting estimated effects.* Information collected from the project proponents, similar projects, census information, secondary sources, and field research can be used to predict future project impacts and compare alternative social outcomes. The baseline social information collected in step 4 provides a starting point for projection. Conditions are predicted for points in the future given that the project (or alternative) goes ahead, and are compared with the conditions predicted without the project (or with another alternative). The difference between the two scenarios compared are the impacts of the project.

   A number of different projection methods are used, ranging from the subjective to the empirical. These include trend projection (taking an existing trend and extending in a linear fashion over time); population multiplier methods (assessing the future changes in variables such as jobs based on population increases); logical scenario construction (developing a hypothetical future based on information collected); fitted empirical scenario construction (using similar cases to develop a scenario, with expert opinion accounting for the variations between the case in question and others); expert testimony (using experts to present their scenarios and assess implications); and mathematical modelling (formulating premises about the project mathematically and using these and other information to analyse quantitatively the effects of a project).

7. *Predict responses* – this step aims to determine how people will respond to significant impacts and effects resulting from a project. It is desirable to know how people will respond with attitudes and actions. Information from the field components, as well as social science literature, may give some indications as to the likely reactions.

8. *The identification of cumulative and indirect impacts* may also be important, as these may be significant but easy to overlook. These impacts may occur much later, or further away, from the time and place of the project, and can be caused by the effect of the more obvious primary impacts.

9. *Alternatives* – this step acts as a feedback loop in the process, allowing the development of new alternatives based on information collected and analysed. The originally conceived alternatives may not be practical or prudent when viewed with the better social understanding that is obtained through the previous steps. In some cases it may

be necessary to develop more alternatives and proceed with the assessment of these.
10. *Mitigation* of identified impacts may be achieved in three broad ways: through completely avoiding the impact (by designing alternatives including abandoning the project); by minimising, rectifying or reducing the impact (through design, operation or policy alteration); or by compensation (based on the severity of the impact or on losses).

    It is at this step that the path to the best overall outcome must be developed. The importance of the significant impacts identified can be ranked, and the ability of the project to mitigate for these can be realistically assessed.
11. *Monitoring programs* to check for deviations from the proposed actions, and to identify any impacts that were not predicted are vital. Monitoring can also check projected scenarios against the actual situation. Monitoring will of course be of little use unless project design and management is such that reaction to the results of monitoring is possible, and mitigation of unforeseen effects is achievable where necessary.

Source: Interorganisational Committee on Guidelines and Principles for Social Impact Assessment (1994)

**Technique 1: Surveys and Questionnaires**

Basically speaking, a survey is the process of collecting information. There are, however, different ways of collecting information. One may be through the act of observation. Another involves the active participation and response of people who have information that may be useful. Setting aside the solely observational form of information collection for the time being, we will concentrate on the collection of information from people. For this type of collection, a choice can be made between methods – those that use a structured form of obtaining information (survey instruments) or those that do not (unstructured interviews, in-depth interviews). In this section we will deal with the use of survey instruments (questionnaires), as a practical form of relatively rapid and cost-effective information collection.

Successful survey methodologies are based on the theory of social exchange. This theory suggests that the way people act depends on the return that their actions are expected to bring. In the context of using surveys, this means that if a person feels they will get more benefit from divulging requested information than it will cost them to do so, they will give that information. Dillman (1978), in *Mail and telephone surveys: the total design method*, applies the theory of social exchange to the survey process, developing a practical and effective procedure for conducting surveys.

Most of the strategies developed for using surveys have evolved in the context of western places, with infrastructure in place such as telephone and mail systems that allow surveys to be cheaply and thoroughly conducted. It is clear that assumptions cannot be made about the presence of such infrastructure in many places. It should in some cases, however, be possible to conduct meaningful surveys in these locations (as many development researchers and practitioners attempt to do regularly), especially if surveys are designed and conducted with reference to the theoretical bases of social science in general and surveys in particular.

There are more subtle problems that may occur when using survey instruments (and that may be particularly pertinent to non-western contexts). Some people may be tired or suspicious of surveys, and afraid of committing themselves by answering formal questions. Byers (1994) discusses situations where surveys have raised false expectations and hopes (as some questions have not been perceived by respondents as wholly hypothetical), and where local people are so accustomed to surveys that they feel it is in their best interests to answer strategically – often to get rid of researchers quickly. In other cases, hostility towards researchers is a problem (potentially a problem using any social research technique), which may be exacerbated by the formal and obviously recorded nature of survey instruments. In such cases, other techniques which allow for a different level of rapport and trust to be developed may be more appropriate. Translation errors may also become part of a survey instrument, and problems may arise when this is not realised until part way through applying the survey. Byers (1994) cites an example of a survey in Tanzania that was abandoned due to problems with developing an instrument that was unambiguous and elicited responses that answered the questions the researchers were trying to ask.

We will not concentrate at great length on the different characteristics of the various methods of applying survey instruments, but will discuss the basic principles of any survey process.

### Identifying the information required

The nature of the information required will dictate many other aspects of the survey process. This stage involves defining what is required (in a general sense, at least) – perhaps the attitudes of local people towards a protected area, information about use of a resource, hopes and aspirations, what problems are perceived, resource knowledge of flora, fauna and systems, perceived solutions to problems, motivations for behaviour, and so on. In some cases, it will not be clearly apparent what information will be available, or will be needed, and a different approach to the research may be necessary, such as preliminary interviews with local people to identify issues and concerns.

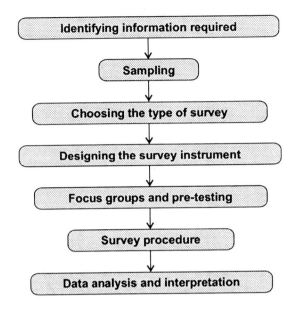

Figure 4.1 The survey process

*Sampling*
Once the type of information needed has been ascertained, it may be important to determine who holds or has access to that information. It is often impractical to survey all of the individuals in possession of information useful to the survey, so a sample is usually taken from the overall group. In some cases, the information required will be held by a distinct group within the society, for example a culturally or geographically distinct group, which will narrow down the number of individuals from which to sample. Sampling is the process of selecting a number of members from the desired group who represent the characteristics of that group. Irrespective of the characteristics of the society or overall group being surveyed, two questions must be answered before sampling can proceed. What size sample is needed, and what type of sample will provide the researcher with the most accurate information?

The size and number of samples will depend on a number of factors, including available resources such as time and funding, the characteristics of the original population, the type of survey being carried out and the statistical accuracy required from the survey. As the sample size increases, so does the cost of the survey procedure and the degree of statistical accuracy of the sample. In many cases, a high degree of statistical representativeness is required of the sample, to allow for extrapolation of the findings of the survey to the overall group, or to allow for effective duplication of the survey in other populations or at other times.

The type of sampling carried out also relies on the characteristics of the original population and the resources available. In some instances random sampling of the entire group may be the most appropriate way to obtain results, which can then be used as a representation of the entire population. Stratified sampling may also be used. This involves dividing the entire group into a number of non-overlapping subgroups from which random samples are taken proportionately. However, these sampling techniques each require the identification of all individuals within the whole population to allow for truly random sampling. Where such information is not available, sampling on a geographical basis, such as a percentage of households, may be necessary. Non-random sampling may also be used, which involves accessing 'key informants' known to be in possession of important knowledge – such as community elders.

The sampling technique used in each survey must be based on the characteristics of the survey requirements, the information available and the characteristics of the overall population. A number of texts, including Bailey (1987), provide an explanation of a number of sampling techniques.

*Choosing the type of survey*
In general, the choice of survey will be specific to the situation and requirements, but needs to be informed. The three main methods discussed in most literature on conducting surveys are mail, telephone or personal interview surveys. In some places, such as remote or rural areas, mail and telephone surveys may not be an option, as the infrastructure necessary for conducting these types of survey may not provide access to those who need to be surveyed. The blending of techniques (such as using personal delivery of a written instrument, to be filled out and later picked up) appears neglected in the literature due to the fact that such approaches are not necessary in the western industrialised city context. Such techniques may be the only practicable way of conducting a survey in many places.

Personal interviewing, as a means of applying a set survey instrument, offers the opportunity to probe for answers to obtain more detailed information on certain topics. In cases where translation problems or misunderstandings are more likely, the personal interview may result in more accurate information being shared. Detailed or complex information requirements may be best satisfied by the use of personal interviews (structured interviewing), in order to explore fully the responses of those being surveyed. Personal interviews can be used in conjunction with visual material, such as pictures, diagrams, maps, cartoons, and so on, to help illustrate what the interviewer may be talking about.

*Designing the survey instrument*
It is important when conducting a survey to ensure that each respondent is asked the same questions in the same way (for those surveys that require statistical accuracy and comparability). If a number of different interviewers

were used, for example, and each asked questions differently (or even different questions) there would be no certainty that different answers from respondents were not prompted by the way that questions were asked, rather than an actual difference in the way respondents felt. For this reason, a survey instrument is usually used to ensure uniformity, in the form of a printed or copied form, booklet or page for a mail survey, or a script for telephone or personal interview surveys.

The design of the survey instrument progresses in a number of stages: writing the questions and additional information; ordering the questions and additional information; and designing the physical layout of the instrument (for mail surveys) or constructing the interview script (for telephone or personal interviews). The literature abounds with advice on question wording, style and ordering, along with other aspects of the survey instrument. Bailey (1987) and Dillman (1978) each provide a useful discussion of the techniques and pitfalls involved in producing survey instruments.

*Pre-testing*

The process of pre-testing helps to refine the survey instrument. According to Dillman (1978), there are three types of pre-testing – colleague pre-testing (asking informed colleagues who are familiar with the study and survey techniques in general to make comments), 'user' pre-testing (where the instrument is commented on by the types of people who are likely to use the results, such as policy makers), and pre-testing with the members of the group being surveyed. A field pre-test, which is in effect a miniature of the actual survey, is desirable to identify problems which may occur under 'real' conditions.

*Survey procedure*

There are, of course, many differences between the ways that survey instruments are applied through mail, personal interview or telephone methods. In Dillman's Total Design Method, it is stressed that underlying the process should be a theme, selected by the researcher according to the survey topic and circumstances, which will ensure that the different elements of the survey process will be consistent and linked in the minds of the respondents. This theme is in the form of a basic appeal, which attempts to convince people that 'a problem exists that is of importance to a group with which they identify, and . . . that their help is needed to solve it' (Dillman 1978: 162). Achieving a high level of participation in the survey is important to ensure the accuracy of the findings. As such, high response rates to mail surveys and high participation levels in other forms of surveys are essential.

*Data analysis and interpretation*

After information has been collected, the process of analysis and interpretation can begin. In most cases, it will be necessary to provide some statistical analysis of the information collected. This may be as simple as providing summaries of the numbers of respondents who answered a question one way or another, or may be more complex. Computer analysis is often used for survey instruments which allow for coded answers. Statistical packages can process survey data, present them in graphical form and carry out analyses of results.

Most social science texts provide some background on statistical analysis, which can act as an entry point for further knowledge. Part of the data analysis stage involves placing interpretations on the results. Care should be taken to ensure that no conclusions are made without results that clearly support them. In some cases, interpretations of results are biased and misrepresent what the results actually mean (intentionally or unintentionally); in the absence of the actual results these interpretations can be very damaging. It is very important to support conclusions with clearly presented results, in order to prove to others that conclusions are valid.

### A Survey of a Community-based Sustainable Farming Movement – Landcare, Australia*

The landcare movement in Australia is a community-based sustainable farming movement, which rests on groups, usually based on a local catchment area, with voluntary membership, attempting to address local and wider sustainable land use and farming issues. Developing since the mid-1980s, landcare is a partnership between government and local communities, where technical expertise and access to funding and government agency staff are combined with the local knowledge, skills and work of local landcare group participants.

The landcare movement in Victoria (a southern Australian state), while appearing to be very successful, had been subjected to very little in the way of evaluation. In 1992, a research project was begun to evaluate the landcare movement across the state. While it was recognised that information from all levels of the movement was important (including individual experiences and broader indications of success) the actual landcare groups were chosen as the most appropriate level at which to access information. The information that was desired was varied, including simple items such as group membership characteristics (numbers, gender and so on), perceptions of the importance of different sustainable farming and conservation issues, goals that the groups had identified, and actions that the group had undertaken to achieve their goals.

The number of landcare groups in the state was approximately 300 when the survey was conducted, and given the resources available for the study no sampling was necessary, as it was possible to conduct the survey as a census, surveying all landcare groups. A mail survey was decided on,

---

* A comprehensive evaluation programme has been applied to the landcare movement since 1992. This case study is based on one of the papers emerging from this programme – Curtis *et al.* (1993). Landcare in Victoria is investigated further in Chapter 9.

as a large amount of information was needed from the landcare groups and some of the information needed to be gathered by the groups for inclusion in their response, which precluded the use of telephone interviews. Personal interviews were considered too expensive, given the wide geographical distribution of the groups. The sponsorship of the survey included the government department that administers the government agency side of the landcare movement. This was considered to increase the likelihood of obtaining a good response rate, as the 'official' nature of the survey would be noted by respondent groups. The survey was conducted according to the Total Design Method (TDM) (Dillman 1978), with cover letters, addressed and stamped return envelopes, reminder card mailouts and follow-up surveys, resulting in a response rate of 53%.

The questions for the survey instrument were based on the fundamental goals of the landcare movement, and the idea that the survey was attempting to measure the success of the movement in terms of these goals. Questions that would provide insight into the achievement of these goals were written. The survey instrument was based on the TDM and was developed in the form of a small booklet of 12 pages, with a number of graphics and pictures. The cover of the booklet used a large graphic representation of a catchment area, and identified the sponsors of the survey and the purpose of the survey.

A focus group, consisting of landcare participants, was convened to refine the survey instrument. Pre-testing was carried out, first in the form of 'user' pre-testing with members of various government agency staff, and then with actual landcare participants.

The information from the returned survey instruments was coded in the normal manner. A number of attitudinal questions had been posed in the survey instrument, which allowed answers along a scale (such as 'To what extent do you agree with the following statement: ... 1 – Strongly agree, 2 – Agree, 3 – Undecided, 4 – Disagree, 5 – Strongly disagree'). Such answers were coded with the appropriate number from that scale, allowing these results to be treated numerically, with statistical advantages such as allowing the calculation of 'average' views about various matters. The coding of such questions along a scale also made it possible to treat them as continuous variables when examining relationships between different attributes. Another component of the analysis involved ranking groups on a collection of attributes (such as number of trees planted) which were considered to be indicative of a successful landcare group, and then adding up these ranks for each group, giving a final score (named the index of effectiveness). The final score of each group could then be compared to the group's other attributes, using correlation analysis, in order to explore what may have helped groups be successful. Some of these correlations revealed, for example, that the greater the percentage of women members in landcare groups, the better the group performed on the index of effectiveness, and also that the higher the level of government funding, the better the performance.

## Technique 2: Interview Techniques

The techniques of interviewing are used in many social science research methodologies. In its most basic form, an interview is a conversational interaction between a researcher and an informant. Rapid rural appraisal, ethnography, social impact assessment, rapid ethnography and other methodologies make use of interviewing as a tool for achieving an understanding of a situation. Different academic disciplines within the social sciences have evolved different names for, and forms of, the interview process. In general, however, interviews can be divided into structured, semistructured or focused, and unstructured interviews (Minichiello *et al.* 1990: 89).

Structured interviews are basically a standardised survey instrument with set questions that are applied orally. As such, they fall into the survey category for the purposes of this book. The survey section deals with survey instruments and the interview method of administering a survey instrument.

Focused or semistructured interviews and unstructured interviews are quite distinct from survey instruments. Focused or semi-structured interviews make use of an interview guide, which contains topics that the researcher has deemed to be of interest before the interview has begun. These topics will usually be those that the researcher feels are pertinent to the social question being examined. The lack of set questions, however, allows the interview to range more widely and the informants to provide their own interpretations and perceptions of the topic. Minichiello *et al.* feel that this form of interview provides a 'more valid explication of the informant's perception of reality' (1990: 92), as opposed to the bias of the researcher being imposed on the respondent in a structured interview (or other survey instruments for that matter) through the presumptions of set questions.

### *Unstructured interviews (in-depth interviews, ethnographic interviews)*

Unstructured interviews are repeated face-to-face encounters between the researcher and informants directed towards understanding informants' perspectives on their lives, experiences or situations as expressed in their own words (Taylor & Bogden cited in Minichiello *et al.* 1990: 93).

Unstructured interviews use a conversational process whereby the researcher learns from the informant and eventually gains an understanding of the informant's perspectives in the context of the latter's own social reality. This is something that cannot be achieved when researchers use their own social context to interpret other people's perceptions.

The important feature of semistructured, in-depth or ethnographic interviews is that they attempt to understand the informant's perceptions of their own social reality. This is accessed by the researcher through conversational interviews, often complemented by the researcher's observations and understanding.

**Plate 4.1** Interview of a herder on the grassland – opportunistic interview held within respondent's own environment. With the stimulus of the grassland environment, and the herder's flocks of sheep and goats, a great deal of information can be obtained. (Photo: R. Thwaites)

Interviews of this nature have a very different conceptual basis to other forms of social research. The varieties of unstructured interviews are used for 'theory-building', rather than 'hypothesis-testing' research. In theory-building research, the researcher does not begin with a view, idea or conclusion about the social issue, but allows the process of conducting and analysing interviews (often in conjunction with observations) to contribute to the building of a social understanding. This process is also known as grounded theory.

While in some contexts interviewing, as part of a wider methodology, can be conducted rapidly, the process of interviewing is generally intensive, in terms of time and the need for prior experience. Spradley's *The ethnographic interview* (1986) provides a comprehensive introduction to all aspects of the process, including the process of learning how to interview. A brief overview of some of the elements of interviewing will be given here but, as for all social science techniques, this book should serve as a starting place for learning about which of these are useful.

*Selecting and approaching informants*
In the context of a particular research problem, a number of informants with different social characteristics may be required to provide the neces-

sary perspectives on the issue. There are some important rules for selecting suitable informants. Informants may need to be thoroughly enculturated (involved or immersed in the culture) which the researcher hopes they will represent. In cases where the research problem involves different cultural groups (or different groups within a culture), a number of different informants from each group will probably be necessary. It is important to recognise that groups may exist within seemingly homogenous cultures – it is easy to define an ethnic group as a culture, for example, but this ignores all of the different subgroups that may exist, and that may be very important to the research issue. For example, some members of that culture may have social status and power through wealth, land ownership or gender, and thus have very different perceptions than members of the culture who have a different social experience.

The researcher also needs to understand how the prospective informant is part of the culture or community being learned about. People in local positions of power may, for example, purport to represent others without that actually being the case.

It may be that the researcher will be unable initially to assess the important distinctions between groups involved in an issue, but will need to look further for other informants, as analysis of early interviews brings distinctions to light. Additionally, prospective informants must also be able to set aside time to participate in the interview process, and need to be aware from the beginning of the amount of time that may be required.

There are differing opinions about the way in which prospective informants should be approached and how much they should be informed about the research. One school of thought suggests that interviews should begin with the informant knowing little about the research question except that the researcher feels they may have some important views which could be discussed. The advantage of this approach is that the informant, knowing very little about the research, will have an uninfluenced understanding of the researcher's interests. Another approach introduces the researcher and the research issue in more detail, telling the informant about the research issue and how the research will be conducted (Minichiello *et al.* 1990).

The decision of whether to discuss all aspects of the research fully with the informant before the interview will depend on the context and situation of the research. In most cases, the second approach is politically and ethically more appropriate (Minichiello *et al.* 1990), ensuring that the informant has full knowledge of what they are involved in, with the first approach reserved for cases that the researcher feels may be particularly subject to bias if the informant knows what the researcher is interested in. Questions of consent and ownership of information can also be more adequately dealt with if the informant agrees to the interviews on an informed basis. Explanations of how the respondent's identity will remain anonymous should also be explicit, if this is going to be the case, before the informant consents to the interviews going ahead. Issues such as remuneration, recording of interviews, and location and timing of interviews may also need to be discussed prior to interviewing.

*Establishing rapport and conducting interviews*
Rapport with another person is defined as 'a matter of understanding their model of the world and communicating your understanding symmetrically' (Minichiello *et al.* 1990: 111), or as 'a harmonious relationship between ethnographer and informant' (Spradley 1986: 78). Spradley examines the issue of rapport from a relationship-oriented perspective, detailing stages of apprehension, exploration, cooperation and participation, which the rapport process goes through. As the relationship moves through these stages, the nature of the information shared will change.

Although in-depth interviews are based on conversational processes, there is substantial choice available for the researcher about how much control over the interview is desired. Many researchers choose to use an interview guide. This basically is a list of themes prepared before an interview, and periodically revised if necessary, that can be used to remind the researcher of what may be important areas to examine. The interview guide can be developed through literature searches and reviews of other work on similar topics, or from prior knowledge or experience. The interview guide does not specify questions, or suggest any order for themes to be discussed, since the 'questions revolve around topics of conversations' as the interview proceeds (Minichiello *et al.* 1990: 115). Other interviewers may prefer a format where themes are addressed as they emerge and the researcher has little control over content and direction.

*Starting interviewing/funnelling*
The early part of an interview process can be directed by funnelling. This technique begins with more general questions, and then directs the conversation towards the issue of interest by focusing the topic via specific questions. This is used to ensure that the conversation is natural and comfortable, before asking directly about the topic of interest. The funnelling technique allows the informant to consider issues at a more general, non-personal level before rapport develops, with more personal issues addressed when rapport, trust and understanding have been established (Minichiello *et al.* 1990).

*Types of questions*
Descriptive questions, structural questions and contrast questions are generally recognised as the main forms of questioning for interviews (Spradley

1986, Minichiello *et al.* 1990). Descriptive questions are where the informant is asked to provide descriptions of people, places, events or experiences (Minichiello *et al.* 1990: 121). These are often used in funnelling.

Structural questions are used to find out about the way informants structure their experiences and knowledge. Spradley recommends their use only after repeated interviews in order to confirm or revise the relationships or information so far discovered. Structural questions attempt to get informants to categorise and locate things in their own words and from their own perspectives. This highlights the points made earlier regarding the interview process. Interviewers, particularly though not exclusively outsiders, cannot assume that informants organise things in the same way that they would.

Contrast questions are the third main type of question used, allowing informants to make comparisons about situations, events, people and so on. Experiences and insights of informants cannot always be understood until we know how they differ, not only from our experiences and expectations, but also from those of other informants. Contrast questions also allow interviewers to understand how informants differentiate and categorise parts of their own experiences.

In addition to these main types of questions, there are a variety of other types that can be used. These include opinion questions, feeling questions, sensory questions, probing or secondary questions, hypothetical questions, devil's advocate questions, and mirroring or summary questions. There are also techniques that are not used in the form of questions, such as expressing ignorance of something, or expressing interest in what the informant is saying in order to encourage them to elaborate. Repetition – of questions and of what the informant has said – is also used to encourage clarification and elaboration.

*Story telling*
Interviewers can also use story telling, which can provide important information about how events occurred (Minichiello *et al.* 1990). A number of techniques exist for prompting a story other than asking outright, including setting beginning and end points of a time period and asking what happened between those points, or by asking questions which imply a process that the informant needs to describe. An interviewer can follow up statements by the respondent by asking for examples to illustrate points that have been made.

*Closing the interview*
There are a number of ways to end the interview process. They will, however, depend a great deal on factors such as the level to which the informant has been involved in the overall research process, the depth of involvement

felt by researcher and informant, whether any actions have been promised by the researcher, and whether it may be necessary to continue interviewing the person at some later stage. The situation may be that contact is never fully broken – the researcher may wish to remain in contact with the informant indefinitely. Regardless of ongoing commitment, the closing of the actual formal interview can be achieved by: explaining the reason for closing; using clearing house questions (such as 'is there anything else that you think may be useful?'); summarising the interview; making personal enquiries and comments, thus moving from an interview situation into a personal conversation; and by expressing thanks and satisfaction (Minichiello *et al.* 1990: 130–32).

**Technique 3: Focus Groups**

Focus groups allow social science researchers to gain an insight into the underlying beliefs and values of particular groups within a community while obtaining detailed information regarding a specified topic. Focus groups achieve this by promoting the discussion of a topic by representatives of targeted groups in the community within the context of issues facing the community at the time (Wynne *et al.* 1993). They are used by researchers when looking into the impact an issue has, or will have, on a community, and on various groups within that community. The information from the focus groups is not intended to be representative of the whole community, but of sections of the community which are deemed by the researcher to have particular relevance to the topic in question (Wynne *et al.* 1993).

In order for focus groups to be a successful technique of qualitative social research, the researcher must first have a basic understanding of the structure of the community. The researcher subjectively chooses various groups within the community to be surveyed for their opinion on the topic. The groups may be (but are not always) collections of individuals with similar characteristics such as socioeconomic status, gender, occupation, geographical location, and so on. Representatives from these groups are then vetted, the choice being based on their ability to represent accurately the group and their existing involvement in the issue. It is best for the individuals to be as similar as possible, and to have no previous involvement or vested interest if a contentious issue is being studied (Macnaughten *et al.* 1995).

A small number of individuals (usually around six to ten) are chosen from each targeted group within the community, and they are invited to participate in the focus groups and offered an incentive to do so. The focus groups then meet individually, with the researcher present as a facilitator, to discuss the topic in question. The researcher introduces various issues and questions to be discussed and allows the groups to discuss them freely and in their own terms. Depending on the research being conducted, there may be a series of sessions at various intervals to allow the participants to develop their ideas and consider their opinions.

The researcher maintains a low profile in the focus group sessions to allow the participants to discuss the topics independently. This discussion usually reveals the social values and the underlying beliefs held by the individuals which are the basis for their opinions towards the topic in question. Thus, focus groups allow the researcher to begin to understand the interactions between various groups within the community and to gain an initial understanding of the cultural basis for a community's reaction to certain issues (Macnaughten *et al.* 1995).

There are essentially two potential weaknesses of the focus group approach as outlined above. The first is concerned with ensuring that the groups targeted are representative of what the researcher/facilitator is trying to understand. For example, targeted groups may be totally irrelevant to the issue or a section of the community which has legitimate concerns which may be ignored or underrepresented.

The second weakness is related to the skills, knowledge and expertise of the facilitator. The facilitator's role is pivotal in the sense that it is up to him or her to guide discussion without imposing values or direction which may be counter to that of the group. A further difficulty in the facilitation process arises from the characteristics of the facilitator (e.g. gender, class, race etc.). These characteristics may impact on discussion and analysis of the issue.

### Wealth Ranking in Southern Zimbabwe*

Focus groups were used in conjunction with quantitative survey methods to determine wealth distribution among farming households in southern Zimbabwe. The research project was designed to analyse the differences in perception of wealth between men and women of the Mazvihwa communal area. Three focus groups were used for the project: a group of local women, a group of men, and representatives of a locally resident research team. Individuals within each focus group were chosen to ensure all socio-economic classes and all geographic areas were represented, and to avoid bias caused by individual characteristics other than gender. The information from the focus groups was compared with the data from a survey of household incomes which had been processed using traditional quantitative research methods.

---

* This case study is based on an analysis of the wealth ranking exercise carried out in southern Zimbabwe by Scoones (1995).

The researchers began the focus group sessions by initiating discussion about the meaning of *mupfumi*, the closest word to 'wealth' in the Shona language. The sessions were held in the Shona language to allow the groups to discuss their opinions freely. The researchers asked the focus group participants to group households from the area into four categories of wealth. These rankings revealed different values between men and women in the community. The men considered that the number of livestock owned by the household determined its wealth, while the women tended to value cash payments from working in the nearby urban centres more highly than livestock. The men argued that households relying on income from work in towns were not wealthy because they always had to buy their food; that is, a wealthy household is one which can provide its own food. The women maintain that households without income from urban industries are at the mercy of unfavourable weather conditions to determine the profitability of their farming practices, and therefore do not have the security brought by a town job. The focus group discussions also revealed an historical aspect of wealth in the society, as various participants reflected that the most wealthy people in their community own fewer livestock than wealthy people did in the past. The differences in the definition of wealth between gender and changes through time reveal the cultural nature of wealth and the value of using focus groups. The raw data collected through the quantitative survey could not reveal trends within society, and cannot give information about the cultural context of what determines wealth in society.

## ECONOMIC ANALYSIS

Environmental economic analysis occurs at a variety of levels. The economic context beyond the project level (international, national, regional or subsectoral) needs to be considered. Munasinghe (1994) states that natural systems tend to cut across the decision-making structure of human society. Because of this, it is important to realise that even if a project at the local level uses sound economic analysis in its design, regional or national policies may influence its effectiveness. Additionally, if the wider economic consequences of a project are not considered, difficulties may be encountered. At the broadest levels, methods such as global environmental economic analysis, environmental accounting and environmental macroeconomic analysis are used. The sectoral, regional, local or project level methods include input–output analysis, extended benefit cost analysis and valuation of environmental impacts.

The use of environmental accounting, for example, occurs at the international or national scale. Conventional indices of national accounts, such as GNP (gross national product) and NNP (net national product) (which are used as an information framework to analyse the performance of economies), use aggregate measures such as savings, investment, consumption and government expenditures. These indices, however, record the productive role of natural resources, whether those resources are renewable or not (so failing to record the depreciation of natural capital), and also fail to record the cost of environmental damages (Munasinghe 1994, and see discussion in Chapter 3). Such measures degrade the environment by ignoring these costs, and therefore provide an inadequate basis for policy formation. The techniques of environmental accounting, which allow environmentally adjusted net domestic product, environmentally adjusted net income and other similar measures to be calculated (Munasinghe 1994), are being developed to provide a better basis for measuring the performance of an economy with environmental considerations taken into account.

Economic analysis at the project level usually means conducting a form of benefit–cost analysis (BCA). BCA is a process, at its simplest, of comparing the benefits and the costs of an option (expressed in monetary terms) and finding out if the benefits outweigh the costs. Prior to environmental economics, BCA included only values that had a market expression (price), but more recently, extended benefit–cost analysis (EBCA) is used to include values that have no potential or present market expression (non-market values, for example the existence of a protected area). There are a number of ways in which EBCA can be used. Alternatives can be ranked, in order to establish priorities, and projects (or policies) can be rejected or redesigned where the costs outweigh the benefits. The EBCA framework can also provide an information system to answer questions about public welfare, or act as an information base for evaluating proposals in the context of regional economics, economic sectors or classes.

There are a variety of tools that are employed to achieve economic and environmental objectives. Jacobs (1991) makes the point that some of these tools (such as permits, taxes, charges, tariffs etc.) have traditionally been ideologically linked, but that this does not need to be the case. Many authors distinguish between regulatory and economic (or market-based) tools. We will provide here a brief description of the main forms of intervention, accepting that the choice of tools is best made according to the desired outcome, regardless of the traditional ideological identification that may be made with that tool.

Regulatory tools (also known as direct control and financial incentive tools) involve actual control, by some level of government, of the economic system. These may include legislative requirements forcing agents to act in an environmentally or socially responsible manner (a law preventing resource extraction in protected areas, for example).

The construction of resource (or property) rights systems can also be employed in order to bring natural resources into the market system. Wildlife, for example, is not normally owned by individuals but by the state or community. Because of this, individual incentives to conserve wildlife are not present. Where ownership or property rights are assigned to locals (either individuals or community groups, as is the case in the CAMPFIRE programme in Zimbabwe (described in Chapter 9), significant improvements can be made in protection.

Although property rights are sometimes complex, creative solutions and systems can be found to ensure that equity and environmental integrity objectives are met. Tradeable permits (for pollution, for example) are another example of an economic tool. Taxes and charges which involve forcing individuals or firms to compensate society for externalities they impose can be levied through the government. Subsidies involving indirect or direct payments to firms or individuals, dependent on their conformation to certain desirable behaviour, can be paid. Tourism operators, formed from local communities that use a protected area in an ecologically sensitive way, could, for example, be offered subsidised access to the resource. Refundable deposits are another tool that can be used to ensure environmental or social objectives are met. While this principle is mainly used in the encouragement of recycling, there is potential for wider application.

## Extended Benefit–Cost Analysis

EBCA can proceed from two perspectives – the decision-making or the social approaches. The decision-making approach suggests that social welfare is multidimensional, with benefit–cost analysis able to discriminate only in terms of economic efficiency, thereby presupposing no ethical principles. The social approach applies the concept of Pareto optimality, enabling choices to be made that will maximise social welfare. Figure 4.2 overviews the steps involved in conducting an EBCA.

### Objectives and scope

The first step involves defining the objectives and scope of the analysis. Important to this step is defining the problem or issue. What do we wish the project to achieve? Who should receive benefits? How far should the effects of the project be investigated (should benefits and costs be considered for a region or wider area)? How much money is available to spend on the analysis? The answers to these questions will help in defining clear objectives for the analysis and for the project itself. Financial objectives may be important, and in conservation/development issues social objectives will also be important. Such objectives may include distributional objectives – to what groups should economic benefit be distributed – among others.

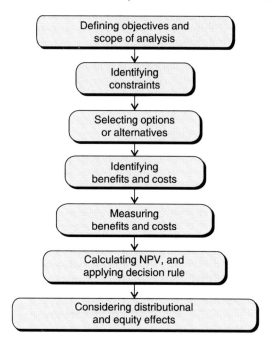

**Figure 4.2** The extended benefit–cost analysis process

*Identification of constraints*

The next stage involves identifying constraints. Under what constraints will the project operate – financial, managerial, environmental or otherwise? This initial establishment of objectives and constraints provides a basis for proceeding with the analysis in a meaningful way.

*Options and alternatives*

The selection of options or alternative ways of achieving objectives is the next step of the process. A number of alternatives may be apparent. One that often needs to be considered is the 'without project' alternative, as the benefits and costs of a project should be compared to what would have happened if the project had not gone ahead. In some cases the only two to be analysed may be the project and the 'without project' alternatives. Sometimes alternatives can be incremental (such as increasing numbers of tourists visiting), but are often not (one cannot build half of a hydroelectric plant, although it can be smaller or larger).

*Identifying benefits and costs*

Identification of the benefits and costs involved in each alternative is the next step. Benefits (values of goods or services, including environmental goods) and costs (foregone benefits) can be very wide ranging, and will include environmental effects. The identification of benefits and costs can be extremely complex, depending on the objectives and scope of the analysis. If, for example, an objective is to achieve maximum social welfare for the whole of society, it will be necessary to take into account the effects of the project on all members of society. Normally, however, the scope of an analysis will be limited to a more practical level. Benefits and costs will differ for different types of project, but for the purposes of introducing the concept, we will provide a brief outline of what may be involved.

Some costs and benefits are easy to identify. Monetary (capital) costs for materials, operation and maintenance, labour, or research and design are usually involved. Similarly, some benefits may be readily identified, such as revenue, avoided costs and reductions in unemployment. Indirect costs and benefits, on the other hand, are difficult to identify and are often not included for practical reasons.

*Measuring benefits and costs*
Measuring benefits and costs, the next step in the process, can be complex. The most straightforward benefits and costs will have market prices, but many will have no market price, or at best distorted prices. When market prices exist, benefit estimates can be based on observed prices. In some cases, elements of natural resource values can be measured where benefits are related to direct resource use (like food or timber extraction) of marketed products. In other cases, for a variety of reasons, it cannot be assumed that prices are accurate indicators of value. For these cases, shadow pricing (determining value by correcting a market price for distortions due to exchange rates or taxation-affected markets) can be used. Opportunity costs may also be used, being the value of alternative choices which are foregone.

Most environmental goods are not traded in markets and so are unpriced or non-market. This, of course, does not mean that there is no value involved – there is just no market expression of whatever value may exist. In the case of a protected area, there may be use and non-use non-market values involved. People may use a resource (for recreation, for example) or wish to retain the option to use the resource in the future. Some use values may be consumptive, others may not. People may also have bequest values for a resource – they value the fact that the resource will be there for future generations to value. Non-market benefits and costs are more difficult to measure, but some techniques are available for doing so.

For areas that are used recreationally, the travel cost method collects information about how much people spend and the distance they travel to reach a place, in order to estimate part of the recreation value of that place. This method becomes problematic when using the place for recreation is not the sole purpose of travel (people may be passing through on holiday, for example), or when congestion at the site artificially reduces demand for that site, which is reflected in the travel cost estimate. A useful summary of the steps involved in applying the travel cost method is provided by Walsh (1986).

Another method for estimating benefits or costs involves the construction of a hypothetical market using the contingent valuation (CV) method. This uses survey techniques to ask people (usually a random sample) what they would be willing to pay to receive a benefit, or what compensation they would be willing to receive to suffer a cost. The CV survey constructs a model market in detail, which is communicated to the chosen respondents. Survey techniques used can include personal interviews, telephone interviews or mail surveys. The hypothetical circumstance in which the good is to be made available – the contingency – is also detailed. A variety of ways exist for eliciting the individual valuations – the most common are stating an amount and asking directly if the respondent would pay, or asking the respondent to nominate a maximum amount they would be willing to pay. Demographic and attitudinal characteristics are also collected to use in analysis and modelling of the information (Mitchell & Carson 1989). The advantage of the CV method includes the fact that it can be used for almost any environmental costs or benefits. CV studies have shown in many cases that otherwise unquantifiable existence and bequest values can be considerable, which is important when considering natural areas and values, as these may often be major components in conservation issues.

*Discounting and applying a decision rule*

An important issue in EBCA is that of discounting. It is argued that society displays a preference for current over future income, and so it is necessary to reduce progressively the value of future costs and benefits. As different benefits and costs will occur at different times in the life of the project, discounting can be used so that benefits and costs are comparable. While most authors argue that a discount rate should be used, some feel that long-term environmental changes should not be subjected to discounting. The selection of an actual discount rate is also very important, as minor variations in the discount rate can cause large variations in the outcome of the analysis.

When all pertinent values have been measured, and the time that they will occur in the future identified, the information can be assessed according to one of three main types of decision rules. One, the benefit–cost ratio, compares the total benefits to the total costs, allowing ranking of project alternatives. A benefit–cost ratio of less than one indicates that economic losses would result from that alternative. Another decision rule, the internal rate of return, expresses the result as the rate of interest which equates the present value of the benefits and costs. This rate can be compared to other financial interest rates to determine how attractive a project is. The other, and generally preferred, rule is net present value (NPV). This rule states that if the sum of discounted benefits exceeds the sum of discounted costs (the NPV is greater than zero) the project is acceptable. The selection of project alternatives with the highest NPV will bring society closest to the optimal solution.

## Distributional and equity effects

None of the decision rules outlined above takes into account equity or distributional effects. If decisions are made purely on the basis of these decision rules, there could be serious implications for the question of who receives benefits and suffers costs. Another concern is that while clearly an extra unit of income is more valuable to a poor person than a rich one, economic efficiency theory, on which the benefit–cost methodology is based, does not allow for consideration of this issue. This means that benefits accruing to lower income groups can be underestimated. Weighting can be applied to the calculation of the benefit–cost outcome, by identifying groups to be advantaged (according to the project's objectives) and assigning weights accordingly.

### An Extended Benefit–Cost Analysis of Protecting Old-growth Forests*

Since the arrival of Europeans on the Australian continent more than 200 years ago, much of the original forest cover has been modified or cleared completely. Old-growth forests (those which have remained essentially unmodified since that time) are therefore uncommon and very important for biodiversity conservation. While some areas of old-growth forest are protected in national parks, some are not, and these areas are under threat from logging. Debate about whether, and to what extent, these areas should be logged has persisted. The EBCA process was applied to this issue to provide decision makers with an economic assessment that included the non-market benefits and costs associated with protection of the forests in national parks.

The main objective of the project was to assess all benefits and costs related to protecting the old-growth forests in national parks (disregarding those known to be of a small enough magnitude not to affect the final result), and to use these values in an extended benefit–cost analysis of that option. The scope of the analysis was defined, as far as possible, by limiting the geographical boundaries of the analysis. Some benefits and costs had already been quantified in previous work, which allowed the project to be further defined. One constraint on the analysis involved the difficulty of valuing some benefits and costs, but these were known to be unlikely to affect the

---

* This case study is based on Lockwood *et al.* (1993).

results of the analysis. Alternatives were identified, which consisted of differing amounts of old-growth forests to be placed in protected areas.

The benefits of protecting the forests in national parks were identified as: the saving of capital and operating expenses associated with logging; preservation of the natural environment; maintenance of recreation and tourism opportunities; reduced damage to roads caused by logging trucks; and maintenance of a forest structure suitable for non-timber uses such as honey production. The costs of protection include: the value of foregone sawlogs; the social and economic costs of increased unemployment; and the loss of recreation opportunities that logging would provide in the form of improved access roads. The majority of these costs and benefits were easily obtained. Non-market values needed to be ascertained for the preservation of the natural environment, and for the non-market component of maintaining recreation and tourism opportunities.

The contingent valuation method was used to measure these benefits, in the form of a survey mailed to 525 members of the Victorian public, selected from the electoral rolls. The survey instrument described and illustrated the location and area of the forests, the forest character and what it looks like, the end uses of logging products, and animals and plant species (especially rare and threatened) reliant on the forests. The people the survey was mailed to were told that the government had to make a decision about protecting these forests. It was made clear that protecting the forests in national parks would cost each household money. For the purposes of the survey, a trust fund was suggested as the way that the money would be collected and administered. An amount of money was stated, and respondents were asked to indicate whether they would be willing to pay that amount. Using the responses, it was possible to sum up the individual willingness to pay values, and obtain a measure of the non-market benefits to society that would accrue from protection.

The total cost of preserving the forests in national parks (only including the socioeconomic costs of unemployment and the value of foregone sawlogs, as the other costs were considered negligible) had a present value of 37 million dollars. The total benefits measured (including saving of logging capital and operating expenses, and the non-market preservation and recreation value) had a present value of 580 million dollars (both benefits and costs discounted over 49 years at 7%). The net present value to society of preserving the forests was therefore 543 million dollars (the benefits minus the costs). Given that the majority of the benefits were non-market, the distribution of benefit from preserving the forests would be very broad, accruing to many individuals in society (as was shown by the individual willingness to pay values collected in the contingent valuation survey). The clear conclusion of the economic analysis was therefore that society would benefit substantially by the inclusion of the old-growth forests of east Gippsland in national parks – the most economically efficient option.

The benefit–cost analysis process can, if applied in its widest manner incorporating the social approach described above, assist in the development of an economically feasible and equitable project. It can be used to understand the distributional and equity effects of a project and its options. The analysis may show that, in order to achieve objectives of equity or to distribute benefits in a way that will benefit conservation of protected areas, using tools of intervention in market systems may be necessary.

## EVALUATION

It is increasingly a requirement of development projects and social change initiatives that their outcomes, both positive and negative, be systematically evaluated during specific stages of their implementation cycle. Evaluation, along with ethnography, has become a buzzword in the 1990s. Everyone seems to be talking about evaluation, and one increasingly meets practitioners from a wide array of social science disciplines who undertake evaluation, either as private consultants, or as part of agency teams.

### What is Evaluation?

So, what is evaluation, how is it done, who can and should do it, and how does it fall into the policy process and facilitate sound conservation and development practices?

Definitions and approaches to evaluation abound, and vary considerably with respect to field (economic, education, health, anthropology), approach (with 'soft', qualitative or ethnographic, versus 'hard' quantitative with pre-test/post-test design features) and range of projects being evaluated (small village level water quality improvement scheme to the impact and outcomes of a large-scale national environment programme such as Landcare).

In addition, evaluation is an emerging and evolving field that cannot be reduced to a set of fixed definitions, recipes or standard practices. Consequently, there are many approaches to evaluation, and considerable disagreement and healthy debate among its growing number of practitioners. Specifically one should keep in mind that evaluation encompasses a 'broad array of questions, purposes, approaches, models, and uses' (Patton 1987: 18). Indeed, as emphasised by John van Willigen (1993: 191), evaluation 'encompasses all the disarray that you would expect in a relatively young field in which persons in many disciplines participate'.

Stated simply, evaluation is 'the process of delineating, obtaining and disseminating information of use in describing or understanding the program, or making judgements or decisions related to the program' (Owen 1993: 7–8). Or, 'evaluation can be defined as any systematic attempt to provide reliable and valid information on the implementation, impact, or effectiveness of a project, program, or policy' (Pillsbury 1984: 44).

'Systematic' and 'reliable' are key words here, as in a sense 'all of us are constantly evaluating things, activities and ideas', but systematic and reliable evaluation must go beyond such 'casually subjective, and largely private, assessments of worth that we produce everyday' (van Willigen 1993: 191). In other words, evaluators must constantly strive to minimise everyday biases and subjective opinions when evaluating a project or programme. This is not always easy, as most people are totally immersed in their own worlds and value positions of what constitutes a 'good' outcome. Similarly, notions of 'objectivity' and 'reliability' may be cultural constructs. Thus, what may be seen as 'reliable' information by a planner in head office may be seen quite differently by a community person on the ground in a rainforest village; what constitutes and is taken as valid knowledge is a matter of perspective and may vary considerably from culture to culture.

**Different Types of Evaluation**

Evaluation, first and foremost, 'is a kind of policy research' (van Willigen 1993: 189); that is, 'policy formation and evaluation' ideally go hand in hand (van Willigen & Dewalt 1985: 1). Namely, 'the purpose of applied research and evaluation is to inform action, enhance decision making, and apply knowledge to solve human and societal problems' (Patton 1990: 11–12).

Similarly, evaluation shares some fundamental features with social impact assessment (discussed earlier in this chapter). Both are concerned with the impact or effects of different actions on people. However, social impact assessment is primarily concerned with discovering *before* the fact any costly unintended effects of an activity, whereas evaluation is most often concerned with determining *after* the fact whether the intended benefits of an activity occurred, or alternatively discovering whether a project with intended benefits is working (van Willigen 1993).

Hence evaluations can function either to improve the functioning of an ongoing project (i.e. a formative evaluation) or to document the outcome of a completed activity so others can use the information (i.e. a summative evaluation). Evaluation may also be involved in monitoring or assessment of progress during implementation.

Evaluation also shares much in common with benefit–cost analysis, described earlier in this chapter. Benefit–cost analysis is just one form of formative evaluation.

Evaluations vary considerably with respect to focus, scope, goals, duration, philosophy, methods and real input into the policy process. Available funding, scope, political and policy context are extremely important considerations governing the process and structure of any evaluation. A limiting feature of most evaluations of development projects is that these 'tend to be short in duration – say, three to six weeks – and informal in methods' (Pillsbury 1984: 53), the point being that it can be quite difficult to investigate the myriad of complex cultural, social and economic issues (not to say the ecological and physical ones) pertaining to the functioning of a programme, in an unfamiliar community, cultural or linguistic setting.

### Evaluation Methods

An extensive literature now exists on evaluation methodology, ranging from high powered statistical analysis, sampling techniques and benefit–cost analysis to very casual, sloppy, 'seat-of-the-pants' evaluation. Pillsbury points out that 'until quite recently, many evaluations of development projects used scarcely any recognizable methodology at all' (Pillsbury 1984: 58). They were in effect 'quick and dirty' or 'windshield' evaluations – 'meaning that the evaluators spent most of their time driving to (and past) project sites with little opportunity to observe actual conditions or talk to either purported beneficiaries or project staff' (Pillsbury 1984: 58).

With respect to the evaluation of development projects, the goal surely should be to develop an adequate and workable methodology between the extremes of the overly sophisticated and the unacceptably sloppy and unsystematic practices formerly associated with development project work. Methodology, of course, is not the only consideration; ethics and value positions come in as well. Specifically, the type of methodology employed will depend in large part on the degree of local level participation and decision-making power built into the evaluation project.

Evaluations today are increasingly employing a variety of methods, both quantitative and qualitative, and often interdisciplinary research terms are used. This reflects a 'contingency theory of evaluation . . . where the legitimacy of a method or concept depends on the circumstances' (Shadish *et al.* 1991: 315). Patton (1982: 311) specifically advised evaluators to recognise the variations in and complexity of the field of evaluation 'having sufficient flexibility to understand which definitions of evaluation are appropriate and meaningful in a particular context, with a specific group of people'.

Yet, many evaluators today rely heavily on a qualitative or ethnographic approaches, especially in rural development settings where it is important to have considerable face-to-face interaction with people in a variety of settings and circumstances. As Patton explained (1990: 47) with respect to such approaches in general:

> Qualitative approaches emphasize the importance of getting close to the people and situations being studied in order to personally understand the realities and minutiae of daily life . . . and that . . . many quantitative methodologists fail to ground their findings in personal qualitative understandings.

## Words of Caution

Doing evaluations raises a number of key issues with respect to the themes of this book. Some would argue that evaluation as traditionally practised is antithetical to local level, bottom-up participatory or empowering philosophies, so painstakingly delineated throughout this book. Some evaluations, it is true, bear all the features of a top-down, bureaucratically intitiated and defined exercise conducted by outsiders, who come in for brief periods of time to interrogate and observe local people, assess and judge, leave, and arrive at a verdict as to the effectiveness and validity of a programme. Following positivistic philosophy and methodology, it is often assumed that such outsiders' findings constitute reliable and objective knowledge (which by definition marginalises or invalidates altogether the legitimacy of local ways of knowing).

Local people and grass roots agency staff sometimes even see evaluators as a threat, akin to insurance investigators and efficiency experts who have come in with a head office hidden agenda to rationalise a decrease in programme funding or even ultimate closure. There may be a strong perception that evaluators employed by agencies or government bureaucracies are in the end beholden to them and will uphold their implicit agenda no matter what. This can operate in quite subtle ways. Quarles van Ufford (1993: 138), for example, describes a discrepancy in the language used by a development agency to describe their bottom-up approach and the on-the-ground reality as observed by agency employed researchers. The anthropologists, in fact, observed a 'virtual absence of any such involvement'. Evaluators in such instances may feel a subtle (and sometimes not so subtle) pressure to report in such a way as to uphold the idealised rhetoric of agency pronouncements. Local people are sometimes also sceptical and cynical about what outside evaluators can accomplish in such short periods of time. Consequently, evaluators are sometimes met with diffidence, suspicion, resistance or outright hostility.

But other more user-friendly models and modes of doing evaluation have emerged over the past few years. Guba & Lincoln's model of 'Fourth Generation' evaluation is certainly a step in the right direction as it recognises in all planning and practice the priority of *all* stakeholders' 'claims, concerns and issues'. It is also meant to be 'collaborative and empowering'. They hold strongly as well to the principle that 'neither the sponsor's nor the evaluator's perceptions are necessarily privileged, but to be taken as "benchmarks" or "constructions"' (Guba & Lincoln 1989: 40). This type of approach serves as a considerable antidote to the more traditional, heavy-handed practice of evaluation.

An even greater privileging of the local level has been developed by Burkey (1993: 125), who argues that 'evaluations must be viewed as an ongoing process of self-evaluation in which the main actors play the dominant role and are the subjects rather than merely the objects of the process'. Only in this way, he argues, will 'genuinely participatory processes' emerge.

**Plate 4.2** Conducting ranger evaluation in Cape York. Informal discussion group with community elders at Injinoo. (Photo: J. Birckhead)

To illustrate the evaluation process in action, the brief overview of an evaluation of the community ranger programme on Cape York in far north Queensland, Australia is presented below. While this evaluation was not able to realise fully the ideal goals presented by Burkey, it attempted to follow some of the approaches derived from 'Fourth Generation' evaluation. This evaluation also incorporated a number of approaches from ethnography, key informant interviewing, informal group interviews and analysis of documents. It is worth considering also that many 'time compromised' evaluations in development contexts employ a number of rapid approaches, especially rapid and participatory rural appraisal. This type of work is usually done by teams of researchers to maximise the saturation and coverage of a research site as systematically and completely as possible in a limited time frame, and often operating within a limited budget.

Pillsbury (1984: 59–60) briefly outlines the ideal activities of such an evaluation team. The case overview in the box includes all of these steps, but the detail and specificity of looking at an actual on-the-ground example should make clearer what one actually does in doing an evaluation:

1. Review of sector-specific design and evaluation guidelines.
2. Thorough review and analysis of documents in project files.
3. Review of all monitoring and evaluation data that may have been collected by project personnel.
4. Review of other project and programme documentation relevant to the project (e.g. sector-specific policy progress of the implementing organisation).
5. Review of published literature related to the subject.
6. Familiarisation with evaluation policy and guidelines of the implementing organisation/s.
7. Interviews with project managers and other representatives of all organisations involved in implementation of the project.
8. Interviews with intended beneficiaries of the project.
9. Interviews with persons not involved in the project (either as implementation or beneficiaries) but who have knowledge of it.
10. Other data collected on visits to the project site/s.

An evaluation of the national Landcare programme in Australia is described in Chapter 8.

### 'Reading' Ranger: an Evaluation of the Community Ranger Programme*

In 1992 Jim Birckhead and Arnold Wallis conducted an evaluation of the Cairns College of Technical and Further Education community ranger programme on the Cape York Peninsula in far north Queensland, Australia. The evaluators were confronted with the task of getting to know a complex, dynamic, many sited programme covering a large and farflung geographic area, in a relatively short period of time. A key feature of the evaluation is that it took place largely within an Aboriginal and Torres Strait Islander (A & TSI) cultural and social milieu; its goal being to tap into A & TSI perceptions of the programme's relevance to their needs; its outcomes and deficiencies. To ensure an A & TSI voice and perspective in the evaluation, co-evaluator Arnold Wallis was chosen for his Aboriginality, as well as his experience in nature conservation and indigenous land management issues.

Loosely following the approach of 'Fourth Generation' evaluation, the evaluators engaged in a type of 'responsive focusing' to privilege the claims, concerns and issues of all stakeholders, as a basic organising prin-

---

*This case study is based on an evaluation programme undertaken by Jim Birckhead in collaboration with Arnold Wallace (Birckhead & Wallis 1994). Other aspects of this case study are covered in Chapter 5.

ciple. Indeed, the approach ideally is empowering, collaborative and educative. The evaluators, as orchestrators of a negotiation process, attempted to identify all cultural reference groups and to tap people's wisdom about their own values and interests by listening to their 'stories', the way they talked about their world, needs and aspirations (Wadsworth 1991).

To achieve these ideal ends within a brief 25 days allocated for field visits by the sponsors, the evaluators proceeded as follows.

1. The programme's implicit theory or logic was arrived at by carefully and critically reading all official policy documents, original proposal and implementation document, various reports, financial and planning statements, correspondence, field workers' reports, press releases, minutes of steering committee meetings and so forth, to establish big picture themes and possible contradictions that hinder progress. From this second-hand picture of the programme, the evaluators were able to 'map' out a working overview of the programme's implicit theory or logic.
2. Ten communities where the programme operated were visited around the Cape. Visits were of one to three days' duration. The format at communities varied between formal structured meetings with community council officials and informal meetings with elders, rangers themselves, other interested community people and participation in field activities with rangers.
3. A number of meetings were held with college staff in Cairns, and group analyses undertaken of their perceived strengths, weaknesses, opportunities and threats (SWOT). In addition, residential schools were attended and course delivery was closely observed. An interesting insight into the programme was also achieved by attending a graduation ceremony and observing the wider involvement of politicians in this event.
4. The evaluators held a series of meetings with government officials and agencies in Cairns and surrounding areas as well as with a number of Aboriginal organisations, and other interested conservation and employment bodies.
5. Finally, collating and interpreting all of the information gathered, including some 30 hours of tape recordings (388 pages of transcriptions) and 250 pages of fieldnotes was undertaken. From this, a draft 'Summary and Recommendations' document was developed and presented in draft form to the consultative committee, representing the various stakeholders, in a draft form to allow further input from this group. Feedback was requested and received, and these additional 'claims, concerns and issues' were incorporated into the Final Report (see Birckhead and Wallis 1994).

> While highly interactive and responsive to inputs from all stakeholders, the evaluation operated within constraints of time and budget, and a very tightly mapped out programme for the evaluators to follow. This was established by the sponsors and the wider consultative committee which represented most of the stakeholders involved. However, the evaluators were constantly on guard not to give greater credence to the sponsor's 'claims, concerns and issues'. These were treated as 'benchmarks' and 'constructions', along with a host of other stakeholder claims and constructions (Guba & Lincoln 1989). Similarly, the evaluators were constantly cognisant of the fact that evaluation, including the evaluators, does not exist outside, or independent of, the programme being evaluated. It and they indeed become another input or event in the programme's history. As Guba and Lincoln (1989: 184) note, 'the evaluation process itself contributes to changes that the monitoring mechanism will detect'.

## CONCLUSION

The methodologies and techniques of social science addressed above have a common theme. They are all employed in various ways and at various times to try to understand the cultural, social, economic and political dimension to human activity. What has also been raised through our discussion is the existence of another dimension to the quest for social understanding, and this is perhaps the most important dimension to the social research process. What we are referring to is how people interpret social information.

The interpretation of social information is crucial, because it is at this point that the paradigm within which we are working more explicitly impacts on the research process. If it is the methodologies and techniques which provide us with the raw material for integrating conservation and development through protected areas, it is the perspectives that we use to inform our interpretations which provide the basis for project development and for social action. To go back to Giddens' (1989) statement concerning the human authorship of both society and change, we, in partnership with local actors, can and do make our own history.

However, we do not always make it as we please. The following chapters deal in greater detail with both the constraints to and avenues for social action and social change within the specific context of local development through protected areas.

# 5 Institution Building and Community Consultation

Understanding the social institutions which constitute a community is crucial if we wish to facilitate a community-based or local level process to promote sustainable development. It is also crucial to understand the processes involved in effective community consultation if we are to achieve desired changes. In this chapter we look at the processes of institution building and strengthening in a community, as well as issues and methods involved in consulting with that community.

## WHAT ARE SOCIAL INSTITUTIONS?

In a social science sense, institutions are the:

> basic modes of social activity followed by the majority of members of a given society... Institutions form the 'bedrock' of a society, because they represent relatively fixed modes of behaviour which endure over time. (Giddens 1992: 731)

There are a number of common institutions which we often think of: the family; the political system; the economic system; religion; and the education system, to name a few.

Institutions are important to our social understanding and to integrating conservation and development because they represent both social entities which are enduring, and a means by which the social, economic and political life of communities, regions and countries are integrated. As we established earlier, integrating conservation and development depends on intervention in existing societal processes, and, often, the struggle for social change. Institutions, as the bedrock or perhaps the glue of society, are an important dimension in these efforts.

For example, if development and conservation need to be made more compatible, we are actually saying that existing social, economic and political arrangements are not compatible (hence the need for intervention in the development process by social change agents, protected area managers and so on). They may not be compatible because they do not protect biodiversity, for example, or because they do not promote democratic or participatory processes which then leads to unsustainable use of natural resources.

What this means is that certain institutional arrangements found within a society or a community are incompatible with integrated conservation and development goals. Logically, then, what local level conservation and development require is the building of new institutional arrangements or the strengthening of and/or alteration to existing institutions in order to make them compatible. This, of course, needs to occur within an ethical framework which incorporates notions of local participation, local control and so on.

In some of the development and conservation literature and experience, the term institution is often used to signify institutional arrangements related to public policy and/or the role of governments and their instrumentalities. Thus, institution building and/or institution strengthening becomes equated with various aspects of the political system.

It is important to note, however, that the process of institution building and institution strengthening can be (and often should be) wider than this, and should encompass the broader understanding of social institutions as described in our opening definition. For example, projects could well strengthen the local economic institution (through ecotourism, the development of new credit arrangements and so on), religious values (through, for example, education programmes) or any number of other possibilities. The important point to note is that the process of institution building and strengthening feeds into alterations to the ways in which various aspects of social and community life are currently organised. Throughout this chapter, we will be looking at the process of both institution building and institution strengthening in this broader sense.

## WHAT ARE INSTITUTION BUILDING AND INSTITUTION STRENGTHENING?

The development experience includes debate between those who advocate the building of new institutional arrangements (in order to achieve desired outcomes) and those who argue for an approach which builds on existing institutional arrangements. At the extremes, the advocates of these arguments would have us believe that conservation/development outcomes would not be possible either without the building of a new society or, alternatively, without using existing arrangements to change existing situations incrementally.

It is therefore important to distinguish between institution building and institution strengthening. In this book, we distinguish between the two as follows:

- institution building can be described as the process of developing new institutions which complement the integrated conservation/development approach;

- institutional strengthening can be described as the process by which existing institutional arrangements are built on and given new social, cultural, economic and political legitimacy in order to integrate conservation aims with development goals.

In any situation where integrating conservation and development is desired, it would not be uncommon for both institution strengthening and institution building to be necessary.

## THE MULTILEVEL DIMENSION TO SOCIAL INSTITUTIONS

It is important to highlight the multilevel component to social institutions. Local communities do not exist in isolation from broader social, economic, political or even ecological processes. The individual is part of a household system, which in turn is part of a broader series of levels to nation-state and global. The individual has a life outside the household level as well, through membership of both formal (for example, producer cooperatives) and informal (for example friendship) groups (see Figure 5.1).

All these serve to influence, and be influenced by, the individual and community norms, expectations and so on. It also means that institutional building and strengthening may impact on both the formal institutions which are readily observable (the development of producer cooperatives or credit cooperatives, for example) and the more subtle, informal arrangements which occur and which are less obvious (for example, networking which occurs at markets and so on).

It is therefore fundamental to remember that one of the characteristics of social institutions is their interdependence. As a result, alterations to one are more than likely to cause alterations to others.

If this is not kept in mind, institutional building or strengthening can have any number of side effects through unintended consequences. In one local level rural development project in India, for example, handcrafts were used as a means of diversifying the local economy. A style of wall hanging which was handmade became so popular that production was increasingly mechanised to keep up with demand. While the local economic institution was strengthened (and further integrated into regional and state levels in the process) and economic benefits accrued to some, the alteration to the production process meant that one of the important social functions of hand making the wall hangings was undermined – a means by which older women could pass on education and information to younger women within the family and the village. As a result, the strengthening of the economic institution meant an alteration to the informal education system as well as a change to the characteristics of the family and roles within it (Furze forthcoming).

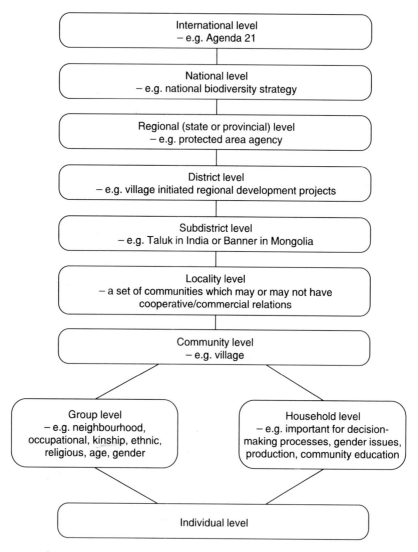

**Figure 5.1** Levels of social, economic and political interaction (adapted from Uphoff 1986)

There are a number of implications which arise from all this. Perhaps the most important is the recognition that institution building/strengthening occurs within both a local level and a broader context. This can result in the building/strengthening process being a balance of local and outside influences and interests.

## Institution Building and Strengthening through Conscientisation

In keeping with the ethical model of local level integrated conservation and development, the ways in which the process of institution building and strengthening can be facilitated are very important. Integral to these is the process of conscientisation.

In his discussion, Stan Burkey (1993: 55) describes conscientisation as a process in which the people try to understand their present situation in terms of the prevailing social, economic and political relationships in which they find themselves'. Therefore:

analysis of reality must be undertaken by the people who decide what their important needs and experiences are, and not by experts. From this analysis, the people themselves may be able to take action against the oppressive elements of their reality . . . [therefore] those who have been considered the objects of development — poor men and women – become active subjects in their own development.

Obviously, the outcomes of this process can be very significant. So too the impacts on existing social, economic and political relationships if these have been based on forms of inequality and exploitation (of both people and nature). According to Dube (1988: 89), conscientisation would:

produce autonomous individuals who would understand, accept and respect their linkages with a wider society . . . They would be reflective individuals, who would at the same time have a participative nature and a positive action orientation. They would be capable of making responsible choices and have the inner strength and self discipline to pursue them resolutely . . . To the extent that meaningful intervention in the process of policy formulation and implementation is essential, they would be politicised.

It is important to realise that the results of conscientisation go beyond the local and affect all aspects of social, economic and political life. It is a fundamental change in awareness and consciousness leading to a renegotiated set of social arrangements as well as relations with nature. In this sense, conscientisation and institution building are complementary, as both social arrangements and the consciousness and awareness of individuals are based on goals of sustainability.

In much development experience, conscientisation has been equated to education. This probably reflects the origins of the term, which came from the important work on education by Paulo Freire (1972) and its popularity in the development/education literature which was critical of the ways in which education became a means of exploitation through teaching uncritical acceptance.

Because of this, difficulties can arise when using [...] opment and participatory approaches. A 'pure' appr[...] sible or desirable within certain institutional arran[...] community or locally based conservation initiatives [...] institutional arrangements both within and outside th[...] may not be receptive or in place. A number of inter[...] be necessary before being able to devolve responsibili[...] development initiatives to the local level. What is imp[...] these interim steps are moving in the correct directio[...] become a mechanism by which the goals are overrun [...]

The balance between future vision, the participat[...] ic necessity is therefore a delicate one and numerous [...] within this chapter and throughout the book. Note t[...] ples use institution building or strengthening as part [...]

### 'Decentralising for Development' through Institutior[...]

In a discussion of alternative development strategies [...] the role of the state and development administratior[...] (1988) suggests that 'imaginative institution building[...]

- *deconcentrating political power from the centre.* [...] tion of political power, and with the resultant ac[...] elites (both urban and rural), the process of deci[...] away from those whom it most directly affects. [...] entrenchment of elites in the political structures [...] concerns which they may well show for structure[...] impact on their ability to rule (as opposed to gov[...] tralisation of political power becomes crucial to [...] This is not to say that with a deconcentrated p[...] place true local participation can always be ach[...] point to, however, is the need to establish a mor[...] native through institution building.
- *reducing the bureaucracy's roles and responsibili[...] limits'*. Planning for development has been chara[...] tion of the planners from those who will be affect[...] An important function of institution building the[...] basis for this separation.
- *enlisting people's decision making in at least thos[...] direct concern to them.* Dube advocates 'minimal s[...] those areas of life that are 'functionally specific'. [...] suggests a political institution which is characterise[...] ing at the local level, by the development of instit[...] local, rather than centralist, characteristics.

## Integrated Conservation and Development through Institution Building and Strengthening

The processes of institution building and institution strengthening are fundamentally important to integrating conservation and development through protected areas. Conscientisation is also important, for it allows true local participation. Conscientisation also acts as a means by which evaluations of local development can take place. We can ask ourselves: has conscientisation taken place?

Before the process of institution strengthening or institution building begins, it is important to have a good understanding of the social context in which protected area management is taking place. This knowledge, developed with local people, should provide the basis for assessment of the current situation and assessing what form of institution building or strengthening needs to occur.

## Institution Building and Strengthening: the Community Ranger Programme at Cape York, Australia*

The community ranger programme which runs in Queensland's Cape York is an excellent example of a local level, grassroots initiative promoting 'bottom-up' development, through training and education of local indigenous people. The programme was initiated by local Aboriginal and Torres Strait Islander communities and established a partnership between government agencies and indigenous groups of Cape York.

In 1987, an Aboriginal Coordinating Council (ACC) held a workshop to discuss management options for Aboriginal land in Cape York. The workshop revealed widespread community support for the establishment of community rangers to help protect, patrol and manage land which had been granted to Aboriginal and Torres Strait Islander (TSI) peoples in Deed-of-Grant-in-Trust (DOGIT), and other land of significance to the indigenous communities. It was determined that the responsibilities of the rangers would include tourism management, land management and legal protection activities.

A second workshop was then held, involving the ACC, members of Aboriginal and Torres Strait Islander communities and Technical and Further Education (TAFE) representatives, to establish an appropriate training programme for the rangers. The outcome of the workshop was a formal request from the Aboriginal and Torres Strait Islander communities that Cairns TAFE establish a suitable training programme, based on the guidelines established at the workshop. Among these guidelines were that:

---

* A recent evaluation of the Cape York Community Ranger Programme forms the basis of this case study (Birckhead and Wallis 1994). Further aspects of this case study are covered in Chapter 4.

- the rangers should be members of local indigenous communities
- transfer of indigenous knowledge should be a major component of the training
- the training programme should provide the rangers with sufficient knowledge to carry out the wide range of their responsibilities
- the course should be structured in such a way that individuals could proceed at their own pace
- the time participants spend away from their communities should be minimised

The TAFE course began in 1989, and has developed to offer a suite of qualifications: the Certificate in Aboriginal and TSI Natural and Cultural Management; the Advanced Certificate in Aboriginal and TSI Natural and Cultural Management; and the Associate Diploma in Applied Science (Aboriginal and TSI Natural and Cultural Management). The ranger programme has stimulated interest from a range of groups, including academics, governments, educational institutions, conservation agencies and indigenous groups. One of the most significant factors in the unique character and success of the programme is the involvement of community elders to convey the traditional ecological knowledge of the country to the rangers. This combines with western scientific training, also involved in the course, to provide the rangers with a combination of traditional and scientific knowledge.

While the success of the ranger programme makes it an example throughout the world of indigenous land management initiatives, as well as the coordination of indigenous and western scientific knowledge, there have been a number of setbacks to the programme. These are mainly due to the innovative nature of the ranger programme, which has resulted in a hesitancy among policy makers to accept its value, and a lack of infrastructure and effective coordination.

Since 1989, ongoing communication with various indigenous community councils about the implementation of the programme has dwindled as other issues occupied the time of the ACC, and the workload of simply teaching the course occupied Cairns TAFE. Also, state and federal funding is insufficient to provide satisfactory employment strategies for the graduates of the programme which can leave the graduates without an obvious career path following their training. In addition to this, the on-ground implementation of management measures by the rangers is inhibited by the absence of satisfactory by-laws to back up their control measures. One of the most significant of these problems, lack of coordination, was addressed in response to a review of the programme. A coordinating body was formed to bring together government departments and Aboriginal and Torres Strait Islander groups.

The many advantages of the programme have been recognised as it has been adopted in the Northern Territory, with Alice Springs TAFE planning to offer the ranger training courses. Aboriginal lands in the Northern Territory occupy vast areas, incorporating a wide range of natural features and land management issues, which can benefit from community ranger management. The community ranger concept is favourable to Aboriginal communities as, if effectively implemented, it can provide training and employment for members of their community, along with improving management practices on their land.

Once the ranger programme is operating to its full potential, it will achieve a range of community involvement goals. These include empowerment of indigenous people, facilitating local level decision making, enabling people to 'care for country', providing sustainable development and community management of tourism and other environmental impacts. While the community ranger programme at Cape York has not yet achieved its full potential, it remains an example to the international community of the way indigenous communities can coordinate their efforts and incorporate their ideas into western bureaucratic and administrative systems to achieve their goals. An evaluation of the ranger programme in 1994 concluded that it was 'an overwhelming success', with the potential to become more effective as the government and other key organisations catch up with its innovative nature.

The community ranger programme highlights the potential of institutional building and strengthening approaches. It also highlights the difficulties which these approaches can face, for example, by being too innovative.

All three case studies described in Chapter 9 discuss institution building and strengthening. The Joint Forest Management programme in India, in particular, promotes the importance of institution building (Sarin 1993).

## COMMUNITY CONSULTATION

### Issues in the Consultation Process

The process of community consultation is central to the integrated conservation/development approach in general and specifically to an approach which incorporates institution building and institution strengthening. Unfortunately, the term 'consultation' has taken on a life of its own, with potential meanings ranging from mere tokenism to the full participation of local people in decisions which affect their lives.

Additionally, the notion of consultation can be, and is, problematic for other reasons. The term can imply a complex relationship between insiders and outsiders or, put another way, between local people and external agencies. Consultation can be 'conferring with a person or persons in order to obtain an expression of opinion or advice' (Jonas 1991: 1) and that advice can be used in any number of ways. For example, it can be used to write up a project document which includes a paragraph or so saying local people have been 'consulted', or it can equally be used to develop a project which reflects the wishes of local communities and which succeeds in exemplifying a participatory, integrated conservation/development approach.

Where a project lands in the range from tokenistic to participatory is dependent on any number of factors. The important point to note, however, is that with participatory approaches, any consultation process is a negotiated one which seeks to achieve results in keeping with the ethical and philosophical stand inherent in the approach. This section explores some of the principles of the consultation processes within the participatory framework.

**Why Consult?**

In an integrated conservation/development context, the participation of local communities is essential. The justification for this has been highlighted throughout this book. Therefore, it is up to us to ensure that the consultation process does not fall into the tokenism category, but enhances the capacity of local people in project design, development and implementation – in short, to ensure that local people more fully participate in matters which affect their lives.

Of course, this means that the consultation process must be a legitimate one, which avoids the pitfalls of tokenism. Nursey-Bray and Wallis (1994: 7–8) have raised a number of points which are directly relevant to this in the form of a series of questions which should be considered before the consultation process: Why consult in the first place? Who is consulting whom? For what purpose are the project and the consultation process? What are the rights of those being consulted? What are the benefits of the project to the individuals, groups and communities involved? What are the disadvantages to the project and who will be affected by them? What is the time frame of the consultative process? Are funds ongoing and adequate to achieve the final result? What cultural protocols are being observed?

Within the specific context of the integrated conservation/development approach, these questions are important.

*Why consult in the first place?* Whilst this might be seen as self-explanatory, it is not necessarily the case. Providing an answer to this question gives you an opportunity to assess or reassess the basis for your work in general, and the consultation process in particular. There are a number of levels at which you can answer this. At the more general level, the consultation process is fundamental to a notion of participation which, in itself, is fundamental to an integrated conservation/development approach. At another level, however, the question also serves to highlight the potential for conflicts between a participatory model and the dictates of policy, funding agencies and organisational structures, to name just a few. On the one hand, then, you may have an ethical basis for a consultation process, yet on the other, you may have external constraints on how much of the information, opinion, insights, needs and desires of local people can be incorporated into a project. As a result, you have a commitment to participatory models, but the commitment may be restricted in its outcomes. Asking the question 'why consult in the first place?' provides an opportunity to reflect on this potential conflict, and to consider ways in which it can be resolved.

*Who is consulting whom?* This question raises some important issues. Who is undertaking the consultation process? Is it only outsiders or are local people involved as part of the consultation team? Is the outsider seen as an expert (so therefore, potentially, even more an outsider)? How has the outsider established rapport with the community? A separate set of questions emerge when we consider who is being consulted as part of the local community, for this raises many issues which we have covered in earlier parts of the book. For example, we have emphasised that local communities are not homogenous, and that a participatory approach means that the rights of the poor, the less powerful and the sick are equally represented. Therefore, whom we seek opinions from becomes vital to the consultation process. Further, how do we get access to a cross-section of the community, to ensure that a meaningful consultation process occurs?

*For what purpose are the project and the consultation process?* Once again, this seems a basic question to ask. However, understanding the purpose of the project is crucial to understanding the purpose of the consultation process. If the project is one which has been developed and imposed by outsiders, then the consultation process becomes one of selling the project to a local population. If, on the other hand, the project has been developed by outsiders in partnership with local people, the purpose of the consultation process is quite different, because it is part of project feasibility, project design, project development and project implementation (not just the implementation phase). Obviously, a participatory approach demands the latter process.

*What are the rights of those being consulted?* A participatory approach cannot see the answer to this question in any other way than rights associated with a capacity to participate more fully and control decisions affecting the lives of local populations. The rights of those being consulted are therefore intrinsically entwined with the notion of self-reliance and participation. If local people, representing all local community groups, are not part of the consultation process and equal partners in project development, then the notion of rights is ignored.

*What are the benefits of the project to the individuals, groups and communities involved?* Answering this question is important in assessing projects which are not the outcome of a community consultation process, as well as in monitoring the outcomes of the consultation process itself. All too often there is an assumption that project benefits are self-evident. It is obvious, for example, that it is in the best interests of individuals, groups and communities that a project which protects certain plant species is a good thing. It may be obvious if this project has emerged from a participatory consultation process. (It may only be obvious to some members of the local community as well.) It may not be as obvious if the project has emerged from outsiders assessing the needs and wants of local people without their input. It may be even less obvious if the project is imposed on local populations and a consultation process occurs in order to sell its merits.

*What are the disadvantages to the project and who will be affected by them?* In our search for projects which benefit local populations in a development and conservation sense, we often forget that projects will have disadvantages. That these disadvantages have to be weighed up with the benefits is obvious. However, it needs to be highlighted that the disadvantages of projects are weighed up during the participatory consultation process, so that local populations are aware of the disadvantages and informed decisions are made. A particularly important part of this is the assessment of what groups the disadvantage will fall on. The conservation/development experience is littered with examples where disadvantages are ignored because they fall disproportionately on the shoulders of the poor, or the less powerful.

*What is the time frame of the consultative process?* A participatory process of community consultation is often a long one. Entry has to be gained, rapport, credibility and mutual trust established and community understanding developed. A potential problem occurs when a consultation process has to be fitted into a project or externally driven timetable. If the desire for a participatory consultation process does not fit within these time frames, then the process may not achieve its desired results. Strategies need to be put into place which ensure that the best possible outcome occurs. It is also worth noting here that the outsider's notion of time may be different to that of local people. Time is a cultural construct and therefore subject to different interpretations cross-culturally.

# INSTITUTION BUILDING AND COMMUNITY CONSULTATION

**Plate 5.1** Gathering in the street in a village in Xilingol Biosphere Reserve. It is at such informal gatherings in public places, here a bus stop, that much information is exchanged. (Photo: R. Thwaites)

*Are funds ongoing and adequate to achieve the final result?* A participatory consultation process raises certain expectations concerning outcomes, possible benefits and processes of project implementation. It is therefore crucial that these expectations can be met within the constraints (financial and otherwise) of projects.

*What cultural protocols are being observed?* If the process of consultation is a process of negotiating power relations, it is also a process of negotiating cross-cultural understanding. Respect for local cultural traditions and the incorporation of cultural perspectives in project design and development are self-explanatory, and highlight the need for a participatory approach to the consultative process (for example Birckhead *et al.* 1996).

## The Process of Consultation

Robson (1993) has cited a number of principles exemplifying a consultative process seeking to enhance local people's participation in decision making and project development:

- the involvement of participants
- negotiation with those affected
- reporting back of progress
- obtaining authorisation before observing
- obtaining authorisation before examining files etc.
- negotiating descriptions of people's work
- considering accounts of other points of view
- obtaining authorisation before using quotations
- negotiating reports for various levels of release
- accepting responsibility for maintaining confidentiality

These principles reflect what Anderson (1992: 79) has called 'simple common sense, common decency and good manners' but they serve to reinforce the centrality of a partnership between local people and outsiders and/or outside change agents. The following case study represents an approach which has been undertaken within the Makalu-Barun National Park and Conservation Area project in Nepal, and which highlights processes of consultation, institution building and institution strengthening.

### Makalu-Barun National Park and Conservation Area Project*

Following the recognition of the value of the Makalu-Barun area, the Woodlands Mountain Institute formed an agreement with His Majesty's Government of Nepal to establish the Makalu-Barun Conservation Project (MBCP). The project provides guidance for the protection of the Makalu-Barun area, from its establishment in 1988 to the year 2000. The MBCP is a coordinating body which provides the infrastructure for a change in natural resource management philosophy among the local communities. In 1991 the Makalu-Barun National Park and Conservation Area (MBNPCA) was gazetted in response to the recommendations of the MBCP, which signalled the start of work in the study area itself. The organisation of natural resource management, in the form of the gazetted MBNPCA and the MBCP management strategy, was essential before community education and projects could begin. This provides the framework for changing community approaches to natural resource management.

The MBCP is aimed at educating the local people about the impacts of their activities on the natural resources of the area, and at highlighting sustainable alternatives to existing economic activities. Funding is provided for projects to assist the move away from economic reliance on the natur-

---

* The Makalu-Barun National Park and Conservation Area project is well documented in the literature. This case study draws on a number of sources, particularly Nepali *et al.* (1990), Nepali (1992) and primary research by the authors. See also Chapter 1.

al resources of the MBNPCA. Thus, the MBCP is strengthening the educational institution by broadening the topics covered by education activities, and by extending education to adults. The economic institution is also strengthened through diversifying economically viable options for independence of local people.

In order to implement the project, members of the local community are employed by the MBCP as motivators. They are trained in a range of skills, including project design, sustainable agricultural techniques, wildlife management, project management, financial management, crafts development, park management, ecotrekking and participatory rural appraisal. The motivators engage in the on-ground activities required for the practical implementation of the project. They are involved in the education process and work with the communities to establish Village Initiated Projects (VIPs), which are funded by the MBCP on an annual basis. The VIPs are the means by which local people can establish facilities which increase their self-reliance and support the goal of the MBCP. The VIPs are proposed by the local communities, and have immediate visible benefit to those involved in their development. VIPs included works such as improvements to drinking water, trail improvement, small irrigation, school roofing and village plantations.

The local community is also involved in the process of accepting proposals for the VIPs through interactive planning meetings with MBCP staff. The motivators are encouraged to focus on existing institutions within the community, to allow people to feel more familiar with the process. The use of existing institutions also helps to maintain the traditional culture of the communities. Such institutions include: village development areas (formerly *panchayats*), which are a government organisational tier; *kiduk*, which is a traditional institution for distributing money; and *gombas*, which are monasteries from which the *lamas* (priests) have localised influence and may be useful as leaders in development activities.

As a part of the move away from traditional economic activities, entrepreneurial activities are encouraged. These activities can supplement the work of the VIPs, as income-generating activities which provide alternatives to destructive natural resource use and improve the livelihood of poor people. Entrepreneurial activities involving women are particularly important as they are the individuals who do the majority of craft work and household activities. In order to encourage such activity the MBCP carries out a range of activities, including:

- providing training for women and other natural resource users to improve design and marketing skills
- establishing combined production and training centres

- establishing an emporium for promoting and marketing local products and encouraging private retailing enterprises
- developing linkages with outside markets for local natural products
- establishing a revolving credit fund for loans for local income generation

Tourism is highlighted as a major opportunity for entrepreneurial activity, and is promoted through the provision of: local access to credit to improve tourist facilities; assistance in the design of handcrafts to meet tourist demand; provision of training to tourist facility operators in management aspects such as health and sanitation requirements; and training for local guides.

A major focus of the MBCP is conservation of the culture of the area. This is seen as essential to the community in light of the projected tourist influx, as it will allow the community to value their cultural heritage and interact with the tourists on their own terms. Facilitating the maintenance of local cultures also protects a component of the area's unique character, and promotes a stable social structure to cope with the major environmental and economic changes which may occur in the area. Some of the activities to conserve the cultures of the area are:

- providing financial and material support for restoration and protection of religious sites
- translating and distributing education materials in local languages
- identifying means to support the culture of local spiritual and health practitioners
- arranging workshops for local teachers and officials to increase their support for local culture
- collecting and documenting items of local material culture, oral traditions and unpublished texts

## CONCLUSION

Institution building and strengthening represent important dimensions to the process of integrating conservation and development through protected areas. They recognise the fundamentally social nature of this integrated relationship. An awareness of this is central to the task of building or strengthening conservation and development goals within societies and communities. To achieve institution building and strengthening and to facilitate participatory action requires a carefully thought-out process of community consultation, taking great care to ensure application of sensitive cultural protocols.

Section 3

# ISSUES IN LINKING DEVELOPMENT AND CONSERVATION THROUGH PROTECTED AREAS

This section examines three key issues in protected area management which exemplify an integrated approach. The integration of development with conservation ideals can be problematic. Not only do outside change agents have to be aware of community or local level social, economic and political processes, they have to understand them within a context of biodiversity conservation and protected area management.

There are obviously many issues which confront policy makers, managers, conservationists, development professionals and outside change agents in general, and their genesis can be local, national or international. In this section, we have chosen three issues that we consider the most important to explore – rural development, indigenous people and tourism. Each issue represents our ongoing theme to search for sustainability through protected area management, and each represents a variety of insights and experiences which can assist our understanding of the integrated approach. In all these issues we highlight how the understanding of social theory and method leads to a more insightful and subtle analysis of relevant issues.

# 6  Rural Development

Rural development is at the very heart of integrating conservation with local populations. This chapter provides an overview of rural development principles and processes and considers their implications.

The idea of rural development has, over recent years, become one of those catch-all phrases which can mean different things to different people. At its simplest, it can mean induced change to existing rural social, political and economic characteristics, as well as alterations to existing agrarian practices. In this way, rural development becomes both a means and an end. It is a means because induced change implies trying to achieve specific, determined goals; therefore, rural change occurs as a result of rural development. It is an end because it is often viewed as the culmination of specific economic, technological and political processes (see, for example, Harriss 1984, Chambers 1986, Burkey 1993).

Rural development, therefore, is at best problematic. When we are talking about induced change, we are talking from a number of value positions concerning the right to induce change, the ways in which it should or should not be induced and the kinds of goals we wish to achieve. All too often, these questions are answered without reference to the legitimate needs and wants of local people. As mentioned earlier, this sets up an 'us and them' scenario, with often disastrous results. These issues complicate any understanding of rural development, and need to be kept in mind as we progress though this chapter and seek to find out how rural development can best be understood.

## THE IMPORTANCE OF RURAL DEVELOPMENT

The importance of rural development to the process of integrating conservation with development through protected areas can be recognised by considering the conservation implications of three important points:

*1. Rural development draws on a theoretical heritage which has concerned itself with understanding the nature of rural society and rural social change.*

Given that protected areas are nearly always in rural locations, the issue of local populations and their relationship with the protected area must be seen to be a rural development issue. The theoretical heritage of rural development should provide a means by which protected area managers and others involved in local conservation initiatives are able to understand and work within rural development frameworks.

2. *Rural development is a practice which is increasingly drawn on in order to incorporate local people into decisions concerning their own futures (that is, the facilitation of participatory rural development).*

Part of the theoretical heritage of rural development practice has seen the emergence of local or participatory rural development as a legitimate strategy on theoretical, practical and ethical grounds. The history of rural development has been a history of searching for appropriate strategies for inducing agrarian change. One fundamental alteration to rural development theory and practice has been the legitimisation of the empowerment of local people through participatory rural development, in what is called more generally the 'bottom-up' approach.

3. *Rural development is at the heart of issues of access to social and economic equity.*

Access to social and economic equity is fundamental to conservation in general and protected area management in particular. It is important to understand the theoretical and ethical basis of rural development as the search for a more equitable distribution of social and economic benefits for rural people.

## UNDERSTANDING AGRARIAN CHANGE AND RURAL DEVELOPMENT

A chapter such as this cannot hope to do justice to the complexity of the topic of agrarian change and rural development. All that can be attempted is to provide a context within which questions can be raised, and the relationship between local people and protected areas can be further understood. With this in mind, let us look at this crucial area.

According to Harriss (1984) there are three main approaches to understanding agrarian change: systems approaches; decision-making models; and structural/historical approaches. Systems approaches emphasise interdependencies. These usually (though not always) involve the study of the interdependencies of environmental, technological and demographic factors and the analysis of social responses to them. Decision-making models emphasise the ways in which individuals respond to market and other factors in their agrarian practice. Structural/historical approaches attempt to integrate the first two by emphasising the relationships between the individual and the broader environmental, social, economic and political processes.

# RURAL DEVELOPMENT

It is important to differentiate between these approaches, for what they highlight is the multidimensional context of rural development and agrarian change. There are many broad processes which result from global and national social, economic and environmental policies which impact directly and indirectly on rural development and agrarian social change. However, there are also many factors which have a direct bearing on rural development which can only be fully understood by emphasising the importance of the household, or the decision-making process at the individual farmer level. It is possible, perhaps even probable, that to neglect one is to interpret only part of the equation.

## RURAL DEVELOPMENT AS A SOCIAL SCIENCE PRACTICE

In a sense, the above discussion centres on theories relating to rural development and agrarian change. However, rural development is more than theorising. It entails a process of action as well, a means of 'doing' rural development.

Broadly speaking, the different interpretations of rural development can be understood to share a common aim: to intervene in existing social, economic and political agrarian formations to bring about some form of change. It is fairly obvious that different approaches will use different assumptions about the nature of the problem, how it can best be managed and what values are important within this approach. For example, some approaches to rural development have emphasised a top-down, policy orientation where the individual farmers and community members are merely the recipients of some form of development action. Others have emphasised a need for local people to be an integral part of the process. Some have emphasised the benefits of the free market system as a means of development, whilst others have emphasised the need for a more centralist state-run approach. Some have wanted change to occur rapidly through the industrialisation of agriculture, others believe that the best approach is to emphasise the primacy of small holdings so that land ownership is widespread. No matter what the approach, when rural development has been considered, the assumption has been one of intervention to facilitate the achievement of certain goals through certain paths.

Intervention in existing social processes raises for some the spectre of social engineering which, carried to its dark finality, results in some very strict form of social control. This has not been lost on social scientists. Discussions on the role of social science in social intervention often result in the emergence of two broad models, the enlightenment model and the social engineering model (Cernea 1991). According to Cernea, enlightenment:

> counts on the dissemination of sociological knowledge through education . . . [implying] a tortuous, uncertain and slow way to return the benefits of social knowledge to society and influence its progress. Moreover, the enlightenment model postulates the dissemination of findings and conclusions available in academic social science, but does not respond to the needs of operationalizing social knowledge for action purposes. (1991: 29)

There are two important points to consider here. First, this model assumes that the enlightenment of social actors will lead to structural change. For example, through education people will somehow bring about institutional change and, ultimately, a new model of development. Secondly, the approach does not necessarily or specifically have social action as an aim. It is more concerned with some form of trickle-down of knowledge (however that knowledge is derived) than more direct intervention in existing forms of social structures.

Cernea suggests that the growth of social knowledge has been such that it can now be used to democratise the planning process itself and, through this, facilitate the broader participation of local people in this planning process. In short, Cernea calls for the social engineering approach to be used in conjunction with the enlightenment model.

Rossi and Whyte point out that:

> social engineering consists of attempts to use the body of sociological knowledge in the design of policies or institutions to accomplish some purpose. Social engineering can be accomplished for a mission-oriented agency or for some group opposed to the existing organisational structure, or it may be undertaken separately from either . . . When conducted close to the policy-making centres, it is often termed social policy analysis . . . When practiced by groups in opposition to current regimes, social engineering becomes social criticism. (Cited in Cernea 1991: 29)

There are a number of points which emerge here as well. First, rural development approaches must contain elements of both models. They must aim to enlighten as well as engineer, or, if you like, educate as well as create some form of structural change. Secondly, an approach which aims for structural change criticises the forms of existing social arrangements. Therefore, one could expect to see some form of institutional revolt or struggle against the implementation of rural development models. Alternatively, one may find institutional support for evolutionary rather than revolutionary models as institutional actors (for example, NGOs, the state, policy makers) balance the competing interests found within the dominant and alternative development agendas. Thirdly, what may occur is a rural development approach which reflects the often contradictory tensions between local participatory models and those which emphasise a more top-down approach.

The relevance of the above debate is this. As discussed previously, knowledge derived from the social sciences should be used to intervene in the integrated conservation/development process. This knowledge therefore plays a pivotal role in the facilitation and implementation of these development strategies. The question is how best to use this knowledge.

In recent times there have been a number of compelling contributions to an approach to rural development which 'puts people first' (Cernea 1991) or 'puts the last first' (Chambers 1986, see also Burkey 1993, Kruijer 1987). The emphasis of these and other approaches has been to facilitate a process of agrarian change which leads to self-reliant participatory rural development.

The ethical and philosophical rationale for this approach is discussed in the opening chapters of this book, and Chapter 4 discusses participatory and rapid rural appraisals as a methodology. An elaboration of the link between rural poverty, rural development and conservation within this framework provides us with an important understanding of both the participatory approach and its importance to conservation and rural development.

## THE LINK BETWEEN RURAL POVERTY, RURAL DEVELOPMENT AND CONSERVATION

As mentioned above, rural development is a means by which poverty and inequality can be overcome. This, however, is not an easy task, not the least because it is so difficult to understand the root causes of poverty.

According to Burkey (1993: 6–11), explanations of rural poverty can be categorised in the following way.

1. *A lack of modernising tendencies.* People following this argument suggest that poverty is caused by a lack of 'modern' technologies or beliefs. They tend to argue that poverty can be alleviated by the adoption of modern techniques of agriculture and/or the adoption of a 'modern' world view. Essentially, the belief is that if changes can occur away from traditional agricultural practice and world views, the inhibitants to growth (and therefore poverty alleviation) can be overcome. It is important to note that, whilst these beliefs are still around in a variety of guises, there has been a growing amount of literature and experience critical of this approach.
2. *Physical limitations.* Often, suggested causes of poverty are related to natural conditions such as soil quality, weather, natural disasters and so on. These explanations can gloss over an understanding of a household's, community's, country's or region's ability to respond to these

so-called limitations. For example, can governments afford to purchase food, is surplus food available, and can individuals and families afford to buy it? Similarly, an individual's, country's or region's experience of health and illness may be less the result of bad luck and more the result of poor nutrition and being unable to afford eradication programmes or drugs. What we see, in the physical limitations arguments, is perhaps only part of the issue, the other parts being related to political, social and economic factors.

3. *Bureaucratic stifling of development.* The adherents of this argument suggest that poverty can be caused by overcentralisation. This leads to programmes which are not only conceived in a distant, urban location away from the 'targets' of the programmes, but often distant in occupational, gender, cultural, social and economic terms. This top-down approach not only stifles programme development but also reduces the poor to being 'targets' of programmes rather than full partners in them.

4. *Dependency of nations.* The adherents of this approach argue that causes of poverty can be traced to global social, economic and political relations between powerful and powerless nations. They argue that this global system of interrelationships is based on exploitation. This occurs where powerful nations have forced powerless nations into a system of dependency, which is based on the powerful becoming richer at the expense of the powerless.

5. *Exploitation of the poor.* This approach suggests that poverty is caused by the exploitation of the poor by local powerful elites such as land holders, local officials, moneylenders, merchants and so on. The continued exploitation of the poor by these groups sets in train socioeconomic conditions where a cycle of poverty becomes impossible to break out of, unless the structures of domination are fundamentally altered. Land tenurial systems which perpetuate inequality are an example of this.

The above categories of explanation highlight the multidimensionality of the problem. Because of this, understanding rural poverty, its causes and the most effective ways to alleviate it is a difficult task. It is, nonetheless, fundamental to an integrated conservation and development framework.

Seeking explanations for causes is one thing, observing the nature of rural poverty within the contexts of these causal explanations is another. Chambers (1986: 13–23), in an important book dealing with the topic, has highlighted a number of biases which may result in poverty being unobserved.

1. *Spatial biases.* Because 'most learning about rural conditions is mediated by vehicles' it is possible that observations and evaluations of rural poverty are only partial in that they ignore the conditions of people away from urban centres, roads and airports.

# RURAL DEVELOPMENT

2. *Project biases*. Chambers suggests that 'Those concerned with rural development and rural research become linked to networks of urban–rural contacts. They are then pointed to those rural places where it is known that something is being done . . . Contact and learning are then with tiny atypical islands of activity which attract repeated and mutually reinforcing attention'. As a result, these projects attract research, funding, the attention of government and non-government organisations and the like, to the possible detriment of other, less well known areas which do not fit into these rural–urban networks.
3. *Person biases*. Local contacts are potentially an important source for bias. Local contacts are used to gain information and understanding, and if these contacts are not representative of the broad range of conditions of rural people, the impressions which we get are likely to be distorted. Some of the more common types of biases which are likely to occur are based on informants being: members of the elite; male; user and adopter (visiting places where activity is easily seen); and active, present and living (the fit, the happy and the strong are more visible than the sick, the weak and the dead).
4. *Dry season biases*. For the majority of rural people, the most difficult time is the wet season (or the depth of winter). However, wet seasons bring with them difficulties of access for the observers and the rural development agencies, so the conditions of rural people are not observed. Ultimately, then, understandings (and project design) are based on dry season observations.
5. *Diplomatic biases*. According to Chambers, diplomatic biases may frequently occur when researchers or project officials avoid asking politically sensitive questions. What may result is understanding which is incomplete because it fails to be based on information from important sources such as the poor, junior officials, and those in conflict with the dominant groups in that society.
6. *Professional biases*. For Chambers, 'professional training, values and interests present problems'. Specialisation of knowledge can lead to a lack of understanding of the interconnectedness of the dimensions to rural poverty. Further, professional training may emphasise a certain world view whereby the professional goes to the field with a collection of assumptions which influence their observations. As a result, these assumptions become self-perpetuating, as they form the framework for data collection.

So, at this stage, we have a number of often competing explanations for the causes of rural poverty. We also have a number of biases which may potentially impact on our understanding of the characteristics of rural poverty which may be within our immediate region. We need to understand this within the framework of participatory local level development and conservation, which we have discussed in previous chapters. What, then, can we do, and how do we go about it?

## IMPLICATIONS FOR PROTECTED AREA MANAGEMENT AND USE

Protected area managers are being required more and more to support the development of local rural communities in and surrounding their reserves. Management practice occurs within a social, economic, political and ecological context. The incorporation of local people into management, and indeed into the practice of conservation, necessarily entails the facilitation of a development process. As emphasised throughout this book, the process must be a participatory one.

However, the defining of participation, the understanding of the dimensions of rural poverty which protected area management may have to attempt to alleviate, and the ways in which rural social, economic and political factors are observed are all problematic. As a consequence, the ways in which these factors are defined, participation is incorporated, and rural poverty is understood and acted upon, will determine integrated conservation and development outcomes. However, as some of the analyses of the causes of rural poverty attest, processes which are instigated well away from the local level can and do impact on it. The potential problems can therefore seem daunting.

The following case study highlights some rural development issues as well as the difficulty of locating these issues within an approach which highlights local participation and an integrated approach to conservation and development.

### The Resource Management Implications of Rural Social Change and Development at Xilingol Biosphere Reserve, China*

Ecological research conducted over a number of years by the Inner Mongolia Grassland Ecosystems Research Station (IMGERS) of the Chinese Academy of Sciences has highlighted a process of biodiversity loss which has been attributed to changing grazing patterns and agricultural practices of local residents of the Xilingol Biosphere Reserve (XBR). The central concern of XBR management is how to understand these changes more fully and ensure that agrarian practices reflect biodiversity conservation.

---

* This case study draws on the experience and knowledge of the authors, based on their research carried out within Xilingol Biosphere Reserve in collaboration with Chinese partners, namely: Baiinxile Livestock Farm; Xilingol Biosphere Reserve Management Bureau; Chinese Academy of Sciences; Chinese MAB Committee; and the University of Inner Mongolia. See also Chapter 10.

This understanding needs to be located within a broader management context which incorporates strategies such as ecotourism, community education and appraisals of management approaches. As a first step, it is necessary to understand the social and economic context of grazing and agrarian change.

This rural development/protected area management component therefore emphasises the production process and changes which are occurring. Social research aimed at developing this understanding has been organised through four research areas: people's perceptions and use of the natural environment; the household; relations between people and production – the village level; and the market economy – household, village and beyond. Each of these represent foci where applied research occurs which is integrated with the central issue of sustainable grazing systems.

The following discussion highlights the research and project development approach which this component has taken, drawing on the rural development experience. It is important to note that the emphasis of the following case is on using social science knowledge to understand more fully a rural development/agrarian social change issue. As such, it provides further application of a rural development approach (both as a theory and as a practice) and also takes on part of the research role of biosphere reserves.

*Why emphasise the production process?*

At the heart of the conservation/development problem in XBR is the changing nature of the production process itself. If, as evidence suggests, Mongolian herders have been grazing their stock at XBR for the last 4-5000 years, and the biodiversity of the grasslands has coevolved with herding, then alterations to the production process itself will place pressure on biodiversity. Whilst this has been pointed out in a number of papers, it has also been highlighted in various ways by a number of informants through interviews and discussions during a rapid rural appraisal. It is, therefore, a crucial area.

Herders and others on Baiinxile Livestock Farm (BLF) comprise a large number of small-scale production units, located within a wider production process. For this reason, it is important to establish social research which occurs at a number of levels (household through to farm and beyond), focuses on linkages within the broader frame of reference (that is, the broader production process) and is coordinated in such a way as to be integrated with management needs, all within a framework of participatory rural development.

*Participatory approaches and the reality of project design and implementation*

The importance of socioeconomic understanding in project design and development has been discussed above. So too has an approach which seeks to incorporate local people, whether they be at the household, local administrative, village or any number of other levels. The importance of this is to provide a means whereby local people, organisations and institutions are able to be meaningful partners in the process of research and project development, not targets of intervention.

Whilst the research subprojects and the applied research areas should not be seen as the final ones in an ongoing process of consultation and discussion, they do represent areas where further information is needed. Hence, and in keeping with the overall design and philosophy of the project, the following points concerning the research were incorporated:

- The collection of data and information should be undertaken as a joint, collaborative exercise with local people, organisations and institutions. The collection of information, after all, is the result of a partnership in the development of social understanding.
- Research and evaluations need to acknowledge that 'local' is not homogenous, and that a variety of 'locals' exist. 'Local' therefore should not be 'tokenism'.
- Project design in implementation, which this research was for, also must occur as a result of the formation of local partnerships.
- These partnerships must be ongoing, both in project development, management, evaluation and monitoring.

These points act as a framework for the research, development, implementation and evaluation phases of the project to ensure that the ways in which it has evolved (through partnerships with academic institutions and reserve management) will not hinder the development of a participatory approach.

*Subproject 1: local people's perception and use of the natural environment*

This subproject gathers information related specifically to the ways in which herders, farmers and others at the local level both understand and use the natural environment. It aims to gather further information relating to cultural bases for beliefs and actions related to the environment, and to contradictions between these beliefs and internal or external pressures for unsustainable land use and the extent of local knowledge dealing with questions of sustainability.

# RURAL DEVELOPMENT

The following research areas have been identified so far.

*Applied research: historical and cultural dimensions to perceptions and uses of the natural environment*   This research collects information related to cultural bases for sustainable grazing. Of specific importance is how potential contradictions between cultural values and herding or agricultural activity are played out. Information on the management of these contradictions should shed light on:

- the cultural values of Mongolian and Han Chinese related to their uses of nature
- how processes of social and cultural change are impacting on these values
- how social and cultural change is impacting on the respective peoples

It should also provide a store of information which will be of specific use to, for example, ecotourism developments and, importantly, the community conservation awareness and education focus of participatory rural development.

*Applied research: local/indigenous knowledge and how this has been incorporated in the broader production process*   This research area builds on the cultural context and emphasises how local and indigenous knowledge of sustainable grazing systems is or can be incorporated into the goals of the biosphere reserve. Is there a contradiction between local and indigenous knowledge and current land use? If so, how has this come about (a question which is emphasised in other areas)? How can indigenous knowledge be reincorporated into the production process? These are some of the questions this research area will provide material to assist in answering.

*Applied research: seasonal land use activity*   Research needs to be conducted on seasonal land use activity from the perspective of 'everyday life'. This would provide a more complete picture on how different seasons in the production process relate, and how these specifically relate to household and village activity.

*Subproject 2: the household*

The household represents the smallest component in the production process. As such, the linkages it has with the village, the BLF, XBR and the wider processes of social and economic change are crucial to understand.

**Plate 6.1** Three generations of women in a Mongolian family, outside their one-roomed mud house on the grasslands of Xilingol Biosphere Reserve (Photo: R. Thwaites)

Examining key aspects to household economy and decision making will develop a better understanding of the social and economic processes which affect, and in turn are affected by, the household. Applied research areas will thus both lead to this greater understanding, and provide a further store of socioeconomic knowledge which will be crucial to resource planning and management, conservation education and awareness, the promotion of sustainable land use patterns and the development of ecotourism and the distribution of its economic benefits.

*Applied research: intergenerational patterns of change in household economy*  This research would act as background to, and also inform our understanding of, household economy over time. Depending on what background information is available, research should be able to be carried out to establish how social and economic activities within the household have changed between and within the generations, as well as an overview of the causes of these changes. An important additional focus is to plot these changes in such a way as to develop an understanding of socioeconomic dimensions to the contemporary household economy. These include such items as income and expenditure generating, and bases for exchange (cash and/or barter and/or cooperative arrangements).

*Applied research: household decision-making processes*  An understanding of the economic decision-making process at household level is crucial to understanding factors impacting on natural resource use, the potential conflicts between natural resource and sustainable land use, and the ways in which these decision-making processes can be used in the context of the development or enhancement of sustainable land use.

*Applied research: natural resource use*  This research seeks to understand patterns of natural resource use at the household level, including such things as herding/cultivation, uses of medicinal plants, household food consumption and so on. Essentially, this is used to build up a picture of the totality of natural resource use at the household level.

*Applied research: diversification of non-herder/non-organic activities at household level*  Little information appears to be available specifically about the amount of economic and social diversification within the household, related to the production system. Interviews have suggested that a process of diversification is occurring, and understanding this is important.

*Subproject 3: relations between people and the production process – the village level*

A central issue in the integration of local people into sustainable development processes (in this case, sustainable natural resource use) is understanding the social, economic and political processes of the community or village levels. If the household represents the smallest component of herding and/or small-scale agricultural production, it is the village or community which represents an important dimension to production wherein social and economic relations between households link with wider systems.

The nature of these linkages highlights a new layer of socioeconomic understanding. This then becomes important specifically for the integration of existing institutional arrangements into systems of sustainable resource use as well as reserve and farm management, and the creation and/or dissemination of community conservation awareness and practice.

Using a community studies approach, this subproject both develops benchmark information of community socioeconomic processes and establishes ongoing, more in-depth analyses of key aspects of community life and community institutions. A number of villages will be selected as case studies to provide this information, on the basis of representing both commonality and difference in relation to the production process.

The applied research areas share a community studies approach but focus on key aspects of village structures. These then represent specific aspects to village life and village structure, which are both important to our overall understanding and crucial to the development and/or maintenance of sustainable resource use.

*Applied research: formal village institutions* Applied social research needs to be undertaken on how the existing, formal institutions found in the villages relate to households, household economy, grazing and agricultural systems and resource use. Whilst the village level administrative structure is one important area, so too is the curriculum in the schools and any other formal institutions which may exist.

*Applied research: social, economic and political structures in villages* This broader research area seeks to develop an understanding of the village as an entity, as well as the nature of its linkages with both the household and the wider economy area, for example the BLF.

*Applied research: informed institutional arrangements/cooperative arrangements in place* Often, some of the more influential avenues for conservation awareness, community education and, indeed, for social life are through informal institutional arrangements which may be in place within a village. These may be anything from discussion groups at the local restaurant to the information communication networks people use to discuss matters of relevance or interest. These are important to understand, as they are often more likely to be the area of 'everyday life'.

*Applied research: the relationship between herding and cultivation* A greater understanding needs to be developed of the socioeconomic relationship between herding and cultivation: of particular potential importance is the role of cultivation in village socioeconomic life, with particular reference to its biodiversity implications.

*Applied research: village level resistance to change* Often, the ways in which a village may resist change, and the reasons for change being resisted, are informative. An understanding of this process often highlights the relationships between formal and informal arrangements (between, for example, the BLF and individual herders), which avenues are best used for

# RURAL DEVELOPMENT

development intervention and community conservation awareness programmes, if/how the stated needs of villages (and individual households) are incorporated into the administrative structures and so on. At present, no information has been collected.

*Subproject 4: the market economy – household, village and beyond*

This subproject specifically deals with the linkages between household and village economies and the broader socioeconomic factors inherent in the production process. In a sense, the subproject represents the final linkage in a chain of understanding which seeks to develop information and knowledge which informs the specific context of the development and maintenance of sustainable grazing systems.

*Applied research: analysis of the relationship between Baiinxile Livestock Farm and the village and household economics* A sophisticated understanding of this relationship needs to be developed. Of particular importance is an understanding of the farm itself, the models of animal husbandry and cultivation which are used, the economic flaws between the farm, village and household and the implications of these, the use of local knowledge in BLF management, the methods of extension used and the perceptions of development used by farm management.

*Applied research: pressures of the urban, industrial economy on natural resource use in the region* The research locates household, village and farm socioeconomic factors within the broader context of urbanisation, industrialisation, privatisation and broader market reform on which China is embarking. These macro level dynamics and policies will specifically impact on the operation of the local socioeconomic system, with implications for the development and/or maintenance of sustainable resource use in the reserve.

## CONCLUSION

Rural development is a significant issue in protected area management. There is a large body of evidence and experience to suggest that the process of rural development must incorporate a conservation and a development agenda and must involve a participatory approach. Such an approach is in keeping with the subject matter of this book, for it highlights the importance of treating development as a conservation issue and conservation as a development issue.

# 7 Indigenous People

Many local or resident people and communities are in fact indigenous or 'first peoples'. Such peoples in the strictest sense, as the terms imply, 'were the original inhabitants of their lands, since colonised by foreigners' (Burger 1990: 16). Known also as 'fourth world' (Burger 1990: 19; Wright 1988: 367) or 'original nations' (Internet 1995b), minority 'nations' within nation-states ('internal colonies') (Bennett & Blundell 1995), there are approximately 300 million indigenous people living in more than 70 countries around the world (Burger 1990; Davis 1993; Kemf 1993). First peoples constitute some '4 percent of the global population' (Burger 1990: 18) and live in diverse environments ranging from arctic and subarctic tundras and forests of northern Europe, Asia and Canada; to deserts and steppes of Africa, South America and Australia; to Pacific islands; and to Asian and South American rainforests. It is estimated that indigenous people 'occupy as much as 19 percent of the world's surface and are, as such, stewards of a significant portion of the earth's fragile ecosystems' (Kemf 1993: 2–3). Their living situations vary as well, from isolated villages and camps in remote environments; to official government settlements, townships and homelands; to urban and suburban enclaves in many countries; and to dispersed housing integrated in their respective dominant societies. Cultural diversity is the hallmark of first peoples. Julian Burger estimates that 'There are some 5000 distinct indigenous peoples in the world that can be distinguished by linguistic and cultural differences, and geographical separation' (1990: 180).

These indigenous communities have one thing in common – their present (or very recent past) close dependence on local ecosystems for their survival. They also share a common impact from a dominant world culture characterised by a high level of national consumption with consequent overexploitation of indigenous people's local ecosystems. The establishment of protected areas in many parts of the world has often resulted in indigenous people being excluded from their local ecosystems and the resources they need for development and even survival.

In this chapter we explore some key issues about indigenous people and their relationship to development, conservation and protected areas.

## WHAT IS INDIGENOUS?

While most scholars would agree on the core features of first or indigenous peoples, the boundaries are actually more open ended and inclusive than suggested in the introduction. As emphasised by Kemf, 'The number of indigenous people surviving today is a matter of definition' (1993: 4). Quoting *The Oxford English Dictionary*, Marsden reflects on the meanings of 'indigenous' – 'born or produced naturally in a land or region; of, pertaining to, or intended for the natives; "native", "vernacular"'. But, as he points out:

> This definition raises more questions than it answers. What is meant by 'native'? What does 'naturally' mean? Is the term equivalent to 'traditional'? An additional meaning is also perhaps implied which refers to 'authenticity' and local 'legitimacy', derived from claims for originality, not in terms of uniqueness so much as in terms of connections with an unbroken historical association with a place. (1994: 42)

How long must the association have been and who has the right to define this? Peoples themselves or social scientists and development workers? As Burger explains:

> First peoples have a strong sense of their own identity as unique peoples, with their own lands, languages, and cultures. They claim the right to define what is meant by indigenous, and to be recognised as such by others. Some now live in cities, earning their livings as, for example, lawyers and community workers – or in many cases struggling to make ends meet; others retain a traditional way of life. But they are united in their desire to maintain their identity and yet be able to adapt and survive. (1990: 16–17)

Smyth, reporting on the IVth IUCN World Congress on National Parks and Protected Areas in Caracas in 1992, notes the discussion which arose 'as to whether the term "indigenous people" had the same meaning throughout the world'. Namely, 'Delegates from African countries . . . expressed the view that the term only had meaning where a colonial power had invaded and subjugated the native peoples (e.g. in Australia, Latin America, North America, parts of Asia etc.). By and large, all the people in Africa are "indigenous people" and therefore it might not be a particularly helpful term in the African context' (1992: 4). Yet, Moringe Parkipuny (Internet 1995c) made a statement before the United Nations Working Group on Indigenous Populations on 'the Indigenous Peoples Rights Question in Africa', arguing that such people include 'those peoples with cultures that are distinctly different from those of the mainstream of national population'. He goes on to say that 'These minorities suffer from the common problems which characterise the plights of indigenous peoples throughout the world. The most fundamental rights to maintain our specific cultural identity and the land that constitutes the foundation of our existence as a people are not respected by the state and fellow citizens who belong to the mainstream population' (Internet 1995c).

Peters similarly presented the case of the Rehoboth Basters of the Republic of Namibia before the 11th Session of the Working Group on Indigenous Populations and the 45th Session of the Subcommission on Prevention of Discrimination and Protection of Minorities of the United Nations Commission on Human Rights. He argued that this group of some 35 000 people are in fact indigenous people on the basis that 'they settled in their lands in the early 1870's' and had 'developed their own legislation years before the Germans installed their colonial rule over Namibia in 1885 and as such they constitute an indigenous people in present-day Namibia' (Internet 1995d).

Marsden sums up the definitional dilemmas:

> Who can we really identify as 'indigenous people'? Are we dealing only with those people who occupy marginal areas – a very small proportion of the human population? Or do we include groups like the Mennonites in the US? How do we deal with the many others who claim rights to separate identity by virtue of their continuous (and original) occupancy of particular tracts of land – the Bretons, the Armenians, the Kurds, the Palestinians? Where can calls for separate identity end? 'Indigenous' begins to refer to an attitude of mind and assumes a struggle for rights somehow abrogated or ignored by a colonising power. It may also refer to those types of organisations that emphasise communal use of resources, untainted with the selfish individualism associated with the expansion of private property. (1994: 43)

Kemf notes that the Worldwatch Institute allows considerable latitude in its estimate of the number of indigenous people. It says 'that roughly 200 to 600 million people are indigenous, taking into account a wide range of varying ethnic groups' (1993: 4).

Similarly confusing is that the notion of 'indigenous knowledge' is sometimes used to include the agricultural knowledge of local farmers or settler cultures. Indigenous knowledge and local knowledge are often used interchangeably. This is also found in some applications of the term 'indigenisation' which refers to 'a bid to get rid of expatriates and is a product of nationalism in the post-colonial era' (Marsden 1994: 44). Seen in this sense, not all colonised peoples who are attempting to oust expatriate regimes are necessarily first or original nation peoples.

We need not get bogged down in definitional issues here, but must note the varieties of uses and meanings that adhere to the term 'indigenous'. In these discussions our main focus is nevertheless on original or first peoples and their relationship to the environment and the land; issues involving sustainability, bio- and cultural diversity with respect to the dual issues of conservation and local level development.

## IINDIGENOUS PEOPLE AND COLONIALISM

Until recently, indigenous people were usually seen by colonisers as 'prehistoric' remnants of passing ways of life, 'people without history' (Wolf 1982), 'primitive' hunter-gatherers or tribal horticulturalists, destined to become extinct to make way for 'civilisation' and 'progress'. Such people, it was further assumed, did not use their lands or resources productively to full economic or technological advantage. 'They' were not really part of the modern scientific world and were thus considered to be not fully human, dispensable with few, if any, rights. Consequently, invasion and dispossession became the lot of indigenous peoples around the world as they were displaced from their lands and resources to open up the country for farms, pastures, mines, logging, cities, military bases, highways, airports, national parks and nuclear testing grounds.

In many countries first peoples were removed from their land to reserves or homelands, geographically separated from the rest of the society, often crowded together with displaced people from other tribes or villages. Such settings usually provided little or no economic opportunities and were rife with social and health ills and human misery. In other countries, indigenous people were pushed to marginal and remote areas not wanted or needed by anyone else (unless, of course, minerals were discovered on such land).

Conventional wisdom usually assumed that such people needed to be developed, modernised and absorbed into the nation-state in one way or another. In reality, most first peoples became bureaucratically administered people, recipients of state welfare and control with few prospects for improving their situations, and in some cases victims of state-sponsored genocide or ethnocide (Keesing 1981, Kuper 1981). Kemf (1993: 4) estimates that 'over the past 150 years, some 30 to 50 million indigenous people have died, including eighty-seven entire indigenous groups in Brazil as well'. Similarly shocking is Davidson's (1994: 194) estimate of 'about two hundred thousand indigenous people a year (still) being killed'. However

> Next to shooting indigenous peoples, the surest way to kill us is to separate us from our part of the Earth. Once separated, we will either perish in body or our minds and spirits will be altered so that we end up mimicking foreign ways, adopt foreign languages, accept foreign thoughts . . . Over time, we lose our identity and . . . eventually die or are crippled as we are stuffed under the name of 'assimilation' into another society (Hayden Burgess, cited in Burger 1990: 122).

As noted by Burger (1990: 11) 'today indigenous people are the most disadvantaged groups in society, suffering the worst health, receiving the least education and among the very poorest'. Indeed:

They inhabit the earth's most fragile ecosystems, where they themselves are among the many species endangered by 'development'. Prospects for their physical and cultural survival correspond to global politics and fluctuations in the world market (Palmer 1989: 12).

## THE DEVELOPMENT OF CONSCIOUSNESS

From their common situation of disenfranchisement, powerlessness, marginality and dependency, indigenous people in the late 1960s began to emerge as an influential force on the world stage. In spite of their small numbers and marginal status, indigenous peoples in recent years have become an influential political and moral force in world opinion and conservation policy. Indigenous struggles against multinational corporations and oppressive nation-states in the name of justice, minority rights and the environment have captured popular imagination, media and scientific attention, and have become a symbolic environmental conscience for an overly exploitative world.

Through political action within their home countries and links with international organisations, indigenous groups in far flung parts of the world have communicated their plight to the wider world and developed enduring organisations and alliances with other such groups around the world. Indigenous organisations and international lobby groups have grown apace (Wilmer 1993).

Julian Burger notes the existence of some '1000 indigenous organisations worldwide' (1990: 138). Such organisations range from small, local village councils, to state and national caucus groups, on to international federations and alliances of indigenous peoples. Notable at the international level is the United Nations Working Group on Indigenous Populations. Begun in the early 1980s, this group's mandate includes: 'Assessing the situation of Indigenous peoples across the globe' and 'drafting principles of international law relevant to Indigenous peoples' (Simpson 1991: 30). The group is presently developing a 'Declaration of the Rights of Indigenous Peoples' which will cover the wide range of concerns and issues faced by first peoples around the globe. Issues considered by the working group include:

> the right of self-determination, autonomy and self-government; treaty rights; the restoration of territorial, land and resource rights; deforestation, depletion of natural resources and other environmental concerns; trans-migration and unemployment; loss of and forced changes in traditional lifestyles; the unremitting consequences of assimilationist policies; preservation of Indigenous languages and cultures; lack of medical treatment and facilities; and other alleged human rights problems, including extra judicial killings and disappearances of Indigenous peoples. (Simpson 1991: 30)

Noteworthy as well is the draft proposal of 'International Covenant On the Rights of Indigenous Peoples' developed by the World Council of Indigenous Peoples. It proclaims indigenous rights in 'self-determination', 'civil and political rights', 'economic rights', including customary subsistence and land use activities, and 'social and cultural rights' (cited in Asch 1984: 132–8).

Other organisations have played pioneering roles in indigenous advocacy. The International Labour Organisation (ILO) (Brennan 1994) and UN Commission on Human Rights, for example, have investigated allegations of ethnocide and genocide in a number of countries (Burger 1990) and have developed principles of indigenous rights and equity in the world community. Groups and coalitions in the 1970s in a number of countries also furthered lobbying and political action on behalf of indigenous peoples and issues. To mention a few of many examples: Cultural Survival, INDIGENA, and the Centre for World Indigenous Studies in the USA; Survival International in the UK; the International Work Group for Indigenous Affairs (IWGIA) in Denmark; Gesellschaft für bedrohte Volker in Germany; the Dutch Workgroup for Indigenous Peoples; Incomindios in Switzerland; and many others, too numerous to cite here (Burger 1990, Wright 1988). In other words, a strong consciousness of what it means to be an indigenous person has emerged worldwide and many networks have been formed in support of indigenous rights. Indeed, 'global indigenous activism' (Wilmer 1993: 23) will no doubt increase in the future, and continues to be an important voice in world politics. With the growth and development of new communications technologies such as the World Wide Web or Internet, many indigenous organisations have sites and bulletin boards on indigenous issues. Such sites provide working papers on indigenous struggles around the world, drafts of legislation and international declarations, research, contact points, and information on forthcoming workshops and conferences.

The United Nations designated 1993 as the 'International Year for the World's Indigenous Peoples'. Its theme, 'Indigenous People, a New Partnership', became a focus for many people and groups worldwide to explore better ways for a variety of government and non-government organisations to relate to and work with the world's first peoples. The eight major concerns of indigenous peoples which provided a focus for discussion, policy and political action were:

- land
- self-government and self-development
- resources
- environment
- culture
- language and education
- health
- social and economic conditions

Although these appear as discrete points or issues, they all are clearly interrelated. The land, the environment and resources are inextricably tied to culture, self-government and self-development, language and education, health, and social and economic conditions.

In addition, recent conferences, workshops and their resulting declarations and principles have articulated issues and preferred practices and legislative frameworks for indigenous–state relationships; usually acknowledging the close connection between first peoples and their environments and the value of indigenous environmental knowledge and land management practices to the conservation of fragile environments. As most first people have been displaced from their lands to make way for development and exist as marginalised minorities within powerful nation-states, 'internal colonies' so to speak (Wright 1988), most such declarations strongly assert the right of indigenous peoples to tenure of traditional lands and rights of resource use. As Maurice F. Strong affirms (Burger 1990: 6):

> In the last decades, indigenous peoples have suffered from the consequences of some of the most destructive aspects of our development. They have been separated from their traditional lands, and ways of life, deprived of their means of livelihood, and forced to fit into societies in which they feel like aliens. They have protested and resisted. Their call is for control over their own lives, the space to live and the freedom to live in their own ways. And it is a call not merely to save their own territories, but the Earth itself.

## INDIGENOUS PEOPLE AND THE ENVIRONMENT

The question of indigenous people and the environment has attracted considerable interest from conservationists, policy makers, the general public and, of course, indigenous people themselves. Recent popular books by Peter Knudtson & David Suzuki (1992) and David Maybury-Lewis (1992) have raised public awareness of their diversity, sophistication, and value to conservation, cultural and biodiversity, and sustainable development.

In our badly degraded and polluted environment, the values and practices of ecosystem people (Dasmann 1976) are appealing indeed, and have much to offer urban industrialised societies. As well, there is a growing international technical and scientific literature (not to mention numerous conferences) on the nature and application of an indigenous land ethic and application and use of traditional ecological knowledge.

Alan Durning (in Kleymeyer 1993: 5) makes a strong case for 'the intrinsic' value of traditional ecological knowledge and practices:

> First, indigenous peoples are the sole guardians of vast, little-disturbed habitats that modern societies depend on more than they may realise – to regulate water cycles, maintain the stability of the climate, and provide valuable plants,

# INDIGENOUS PEOPLE

animals and genes. Their homelands may harbour more endangered plant and animal species than all the world's nature reserves. Second, they possess, in their ecological knowledge, an asset of incalculable value: a map to the biological diversity of the earth on which all life depends. Encoded in indigenous languages, customs, and practices may be as much understanding of nature as is stored in the libraries of modern science.

It was little appreciated in past centuries of exploitation, but is undeniable now, that the world's dominant cultures cannot sustain the earth's ecological health without the aid of the world's endangered cultures. Biological diversity – of paramount importance both to sustaining viable ecosystems and to improving human existence through scientific advances – is inextricably linked to cultural diversity.

**Plate 7.1** Aboriginal elder of Kowanyama Community on Cape York, carving up a freshly roasted agile wallaby. The wallaby was cooked wrapped in paperbark (Melaleuca), and buried under earth in a fire of hot coals. Termite mounds in the background are typical of the landscape of Cape York. (Photo: R. Thwaites)

A similar assessment was presented by the World Commission on Environment and Development report *Our common future* (cited in Davis 1993: 1):

> These communities are the repositories of vast accumulations of traditional knowledge and experience that links humanity with its ancient origins. Their disappearance is a loss for the larger society, which could learn a great deal

from their traditional skills in sustainably managing very complex ecological systems. It is a terrible irony that as formal development reaches more deeply into rain forests, deserts, and other isolated environments, it tends to destroy the only cultures that have proved able to thrive in these environments.

The 'stewardship of the earth' (Davis 1993: 1) ethic of first people and their sacred and cosmological ties to the land have captured the imagination of many conservationists, deep ecologists and managers. These holistic messages hold an appeal to people who customarily treat this land as a commodity to be bought, sold and exploited, regardless of the long-term ecological consequences.

Recent United Nations articles and agendas further specify the points noting the rights of indigenous people and their role in conservation, but roles that allow for economic and cultural development:

> [Indigenous people] have developed over many generations a holistic traditional scientific knowledge of their lands, natural resources and environment. Indigenous people and their communities shall enjoy the full measure of human rights and fundamental freedoms without hindrance or discrimination.
>
> In view of the interrelationship between the natural environment and its sustainable development and the cultural, social, economic and physical well-being of indigenous people, national and international efforts to implement environmentally sound and sustainable development should recognise, accommodate, promote and strengthen the role of indigenous people and their communities. (Part VI, Article 28, of the Draft Declaration as agreed upon by the members of the United Nations Working Group on Indigenous Population at its Eleventh Session)

> Indigenous people have the right to the conservation, restoration and protection of the total environment and the productive capacity of their lands, territories and resources, as well as to assistance for this purpose from States and through international cooperation. (United Nations Conference on Environment and Development, *AGENDA 21: Programme of Action for Sustainable Development 1992*: 227).

Granting of land rights, comanagement of parks and protected areas, eco- and cultural tourism ventures, patents and regulations on pharmaceutical and other intellectual property, wealth and employment schemes all come into the picture. Issues of sustainability and local level 'bottom-up' control are important here.

Davis (1993) poses these key practical questions that help to assess the specific situation of an indigenous group:

- What are the traditional views held by these peoples about land and the environment?
- How have national laws and government policies either corresponded to or conflicted with these views?

- What types of policies, programmes or projects could more adequately take indigenous views of land and the environment into account?

The following case study, on indigenous knowledge systems of the residents of the Cocles/Kekoldi Indian Reserve, Costa Rica, highlights a number of issues linking indigenous people and the environment.

### Indigenous Knowledge Systems: Cocles/Kekoldi Indian Reserve, Costa Rica*

> Don Anibal is a passionate defender of the forest, for it provides him the animals and birds whose skins, bones, organs and feathers are needed for spiritual curative purposes, as well as the medicinal plants he was trained to use. When I suggested to him that sharing his knowledge of the medicinal properties of the forest plants with botanists at the National Museum (without revealing the secrets of his spiritual training) might draw the attention of the national scientific community to the importance of defending the Cocles Reserve's forests, don Anibal agreed to bring the plants he could collect to my house in Puerto Viejo. I passed on to don Anibal the National Museum botanists' instructions for plant collection and pressing procedures; he followed them carefully, and I recorded his descriptions of where the plants are found as well as their proper preparation, doses and expected effects (Palmer 1989: 38–9).

Don Anibal is a Bribri Indian living in the Cocles/Kekoldi Indian Reserve in the Atlantic lowlands of Costa Rica. In 1986 a participatory action research (PAR) programme was implemented in response to the declaration of a wildlife refuge adjoining the reserve. The programme was initiated by two NGOs and was intended to determine the existing and traditional resource management methods of the Bribri people of the reserve. This information was to be used to identify possibilities for managing the reserve in accordance with the objectives of the wildlife refuge.

As the principal researcher engaged by the NGOs, Palmer approached the primary Indian authority in the reserve, the Junta, to begin an ethnographic study. The members of the Junta revealed a range of concerns held by the Bribri. The Junta elected four community representatives to work with Palmer as a task force to implement the research programme, thus changing the ethnographic study to a PAR programme. Among the issues identified by the Junta were: the escalating scarcity or loss of native species in the reserve; the illegal hunting of native species by people from outside the reserve; settlement and clearing activities of non-Indian people on the reserve; poor communication between the three main cultural groups in

---

* This case study is based on the report of extensive participatory action research carried out in 1986–7 within the Cocles/Kekoldi Indian Reserve (Palmer 1989).

the region – Bribri, Afro-Caribbeans and Hispanics; lack of government support for Bribri efforts to address illegal land use within the reserve. Another issue, which emerged from the PAR programme itself, was that the traditional ecological knowledge held by the older members of their community was not being effectively passed on to the younger generations.

The outcomes of the PAR programme, which involved interviewing members of the Bribri Indian community from across the reserve, addressed these issues. The task force produced a book and a slide show, describing the relationship between Indians and the forest, spiritual beliefs of the Bribri and the legal status of the reserve, which were used to educate young Bribri as well as to improve cross-cultural understanding within the region. The slide show was also used to improve communication between the Bribri Indians and government departments, and led to the establishment of a training course by the federal Department of Natural Resources which permitted Bribri representatives to have the authority to control settlement on the reserve, and to fine illegal hunters within the reserve. The coordination of traditional Bribri knowledge and western management methods, facilitated by the PAR programme, has resulted in the establishment of a nursery for cultivating threatened flora species from the reserve's forests, and the introduction of a project for semidomesticating green iguanas, which are a traditional food of the Bribri and have become rare in the reserve in recent years.

The quick recognition by the Bribri Indians of the opportunities offered by the PAR programme allowed them to gain support from the dominant western bureaucratic and political system to protect their reserve from deleterious external influences. Don Anibal exemplifies the approach of the Bribri Indians in their attempts to protect their forest, utilising western management methods to achieve their goals. The strong cultural identity maintained by the Bribri throughout the invasion of non-indigenous settlers, which has more than doubled the regional population since 1980, allows them to utilise western methods without losing their cultural identity. The empowerment of Bribri Indians, which is based on the recognised validity of their traditional ecological knowledge, provides an illustration of the influence indigenous groups can have in protected area management.

## INDIGENOUS PEOPLE, PROTECTED AREAS AND LOCAL DEVELOPMENT

There are approximately 8000 protected areas occupying between 5 and 6% of the world's land surface. Many of these coincide with indigenous people's 'country' and the ecosystems they require to sustain their livelihood and provide for future development. While the first wave of protected areas that were established across the world often alienated indigenous people from their land, in recent times public policy concerning protected areas has undergone a fundamental change towards a genuine partnership between indigenous people and protected area managers.

**Plate 7.2** John Clarke, Aboriginal Ranger at Kowanyama Community, Cape York, shooting feral pigs. This area has been recently burned – Aboriginal people have been using fire as a land management tool for over 40 000 years. (Photo: R. Thwaites)

The IVth World Parks Congress in 1992 endorsed 'The Caracas Declaration – Protected Areas and the Human Future' which enshrined the rights of local and indigenous peoples. The recommendations of Workshop 1.5, 'Indigenous views of protected areas', best encapsulates the progressive principles agreed upon by the majority of participants.

   a. IUCN should recognise indigenous territorial rights as a pre-condition for the management and establishment of protected areas which involve the home lands of indigenous peoples.
   b. IUCN should recognise indigenous environmental knowledge and traditional management systems as the principal basis for protected areas planning, management and use.
   c. IUCN should support the documentation and application of indigenous knowledge and traditional resource management methods.
   d. IUCN should recognise the important role of women's knowledge and participation in protected area planning, management and use.

e. IUCN should recognise the intellectual property rights of indigenous peoples and promote appropriate compensation for the provision of services.
f. These recommendations and future IUCN activities involving indigenous peoples should be carried forward by an inter-commission Task Force on Indigenous Peoples. (IUCN 1993: 85–6).

A number of recent works have addressed these issues with respect to a variety of indigenous cultures around the world, for example Birckhead *et al.* 1993, Freeman & Carbyn 1988, Kemf 1993, Williams & Baines 1993, McNeely & Pitt 1985.

What is becoming clear is the need to find creative ways to link the ecosystem management practices of local indigenous people to the ecosystem management principles of protected area managers. At the same time the local community development needs of indigenous people must be strongly promoted while seeking the wider conservation and development goals of the nation-state.

McNeely (in Kemf 1993: 253) lists 10 principles to achieve these linkages.

1. Build on the foundations of the local culture.
2. Give responsibility to local people.
3. Consider returning ownership of at least some protected areas to indigenous people.
4. Hire local people.
5. Link government development programs with protected areas.
6. Give priority to small-scale local development.
7. Involve local people in preparing management plans.
8. Have the courage to enforce restrictions.
9. Build conservation into the evolving new national cultures.
10. Support diversity as a value.

Richard Chase Smith (1987: 8–12) poses a number of crucial questions regarding grassroots development and autonomy:

1. Is the indigenous community in control of the conceptualisation, planning and implementation of their development?
2. Does the indigenous community exercise control over its territory and over all the resources found within the limits of that territory?
3. Does the program for development promote self-sufficiency and economic independence of the indigenous community?
4. Does the development process strengthen the social and cultural bonds of the community and affirm the sense of historical identity and cultural dignity of the community members?

One of the models which has been emerging that attempts to link directly the community development of local indigenous people with biodiversity conservation through protected areas is joint or comanagement. There are a number of examples of this process and one, the Uluṟu/Kakadu model in Australia, is described in the following case study.

## Joint Management at Uluṟu-Kata Tjuṯa National Park, Australia*

*Handback*

In 1985 the Uluṟu-Kata Tjuṯa National Park was handed back to its traditional owners. The joint management arrangement which has evolved since then between the indigenous Aṉangu people and the federal park agency is now lauded around the world as the Uluṟu/Kakadu model of joint management.

The land now covered by the park was declared part of the Petermann Aboriginal Reserve upon European settlement. However, in 1958 the rights of the Aṉangu – the indigenous population – were revoked by the government when it established the Ayers Rock-Mount Olga National Park on the site. At that time, national parks in Australia followed the 'Yellowstone model', and so the Aṉangu were excluded from inhabiting and utilising the resources of the park. In 1985, after years of extensive lobbying by the Aṉangu and wrangling between the Northern Territory and Commonwealth governments over the conditions of the handback, the national park was returned to Aboriginal ownership.

Prior to handing back Uluṟu-Kata Tjuṯa to its traditional owners, a number of conditions were made, including:

- Aṉangu would have freehold title under Commonwealth legislation;
- immediate leaseback of the land to the Commonwealth government, under the authority of the Australian Nature Conservation Agency (ANCA), to maintain community access to the site;
- a board of management would be established for the national park with a clear Aboriginal majority;
- the lease would detail the obligations of ANCA to the Aboriginal owners along with details of the Aṉangu rights of use and residence in the park;
- an annual rental payment and a portion of park entrance fees would be paid to the Aṉangu by ANCA.

---

* The Uluṟu-Kakadu model of joint management and its implementation at Uluṟu-Kata Tjuṯa National Park are discussed widely in the literature. This case study draws from such literature and in particular De Lacy (1994) and De Lacy & Lawson (1996).

## Lease and board of management

State and commonwealth legislative amendments accompanied the establishment of Aboriginal owned, jointly managed national parks, and provided the key to true power sharing by indigenous people. The commonwealth legislation required that a board of management be established for all Aboriginal owned land consisting wholly or partly of a national park or reserve which would:

- produce a plan of management for the land;
- implement the plan of management in the park or reserve;
- monitor the management of the park or reserve;
- advise the Minister on future developments within the park or reserve;
- carry out its activities in conjunction with the Director of ANCA.

The terms of the lease to ANCA at Uluru-Kata Tjuta also promote Aboriginal interests through a number of additional requirements, such as (Woenne-Green *et al.* 1994: 285–6):

- encouraging the maintenance of Aboriginal tradition within the Park through the protection of areas, sites and matters of significance to the traditional owners;
- taking all practicable steps to promote Aboriginal administration, management and control of the Park, and to urgently implement a programme for training Aboriginal people in skills needed to do so;
- involving as many Aboriginal people as possible in the operations of the Park and adjusting working hours and conditions to the needs and culture of Aboriginal people employed in the Park to facilitate this;
- maximising the use of traditional skills in the management of the Park;
- promoting among non-Aboriginal employees in the Park (and where possible among the visitors to the Park and residents of Yulara) a knowledge and understanding of the traditions, language, culture and skills of Aborigines, and to arrange for proper instruction by Aborigines engaged for that purpose;
- regularly consulting with the traditional owners and their organisations about the administration, management and control of the Park; and
- encouraging Aboriginal business and commercial initiatives and enterprises within the Park.

The implementation of these policies is one of the major roles of the board of management and an integral part of joint management.

The board of management is also responsible for determining park policy, carrying out park planning and involvement in day-to-day management of the park. The Anangu hold a majority on the board of management, six of ten members, which also includes the Director of ANCA, an arid zone ecologist, a nominee of the federal minister responsible for tourism, and a nominee of the federal minister responsible for the environment. Both the lease and the plan of management for Uluru-Kata Tjuta National Park have an in-built review process requiring five-yearly revisions.

*Joint management*

A major consultancy was carried out for the Mutitjulu community in 1989 to assess the progress of joint management, which led to a number of changes to the management structure, many of which are outlined in the 1991 plan of management for the park (Uluru-Kata Tjuta Board of Management and the Australian National Parks and Wildlife Service (ANPWS) 1991). Involvement of Anangu in everyday park management has been a focus of the park since joint management began. A major joint project between CSIRO researchers, park staff and Anangu representatives was carried out in 1990 to conduct a major fauna survey of the park. The survey integrated traditional ecological knowledge and western scientific knowledge and promoted acceptance of traditional ecological knowledge to the scientific community.

Even with all of these provisions, is Uluru-Kata Tjuta truly an 'Aboriginal National Park' (Uluru-Kata Tjuta Board of Management and the ANPWS 1991: 1)? At the base of this question is the nature of Aboriginal ownership of the land. Granting ownership on the condition of immediate leaseback of the land to the government for community access can hardly be held to provide the Anangu with total self-determination. When the significance of 'country' to Aboriginal Australians is considered, land rights in the form of ownership and management control can be seen as essential to their cultural and social survival. Within the model of joint management of Aboriginal owned national parks, the ability of Uluru-Kata Tjuta to be called an 'Aboriginal National Park' can be assessed by addressing a number of issues, including the features of the lease and management structure, and the nature of Anangu involvement which stems from these.

The lease agreement between the Anangu and the Australian government secured the rights of the Anangu to occupy the land of the national park and to utilise its resources. This has been exercised to date by the maintenance of the Mutitjulu community and carrying out subsistence hunting in a remote area of the park. The 1991 plan of management (Uluru-Kata Tjuta Board of Management & ANPWS 1991) states that the management of fire, along with a number of other management issues within the park, will be based as far as possible on traditional knowledge and methods. Traditional fire management methods will be used in an attempt to create an environment favouring the recolonisation of various species which have decreased in numbers since the cessation of Aboriginal land management. However, there are doubts about the security of the lease agreement. The major federal opposition parties have adopted a policy to return management responsibility for Uluru-Kata Tjuta National Park to the Northern Territory government. Such a move would set back joint management in the park as new agreements and management arrangements would need to be made.

Many conditions of the current joint management arrangement have evolved from the early experiences of Anangu involvement at Uluru-Kata Tjuta. The original plan of management, produced in 1986, offered little guidance for implementing the policies of joint management which had been established. The plan provided repeated expressions of goodwill, but little in the way of practical steps by which Aboriginal involvement and empowerment could occur. It soon became obvious that the key to effective joint management was the incorporation of Anangu staff into the everyday activities of the park. One of the most important components of achieving this goal was to establish flexible positions within the ANCA park management team which were suitable for Anangu employment.

Evidence of the effective integration of Anangu and park staff in the management of the park is the involvement of the Mutitjulu community. It is the only community within the park, and is home to many of the traditional owners of the land along with ANCA staff. As aspects of the management of the park have become more and more a part of the daily activity of the Anangu, the distinction between park management and more traditional Mutitjulu community business has become less significant.

*Financial benefits*

The lease between the Anangu and ANCA provides for significant financial benefit to the Anangu community. A recent review of the terms of the lease raised the annual rental for the park to $150,000. In addition to this, the Anangu receive 25% of all park entrance fees and a portion of the income from fines and charges imposed within the park. The Anangu are also in a position to gain financially through providing products and services to tourists visiting the park. Extensive Anangu involvement in the tourist experience at the park promotes a wider understanding of their culture and the Aboriginal importance of Uluru-Kata Tjuta. However, Anangu involvement in tourism activities is affected by their inability to control tourist numbers within the park. The Mutitjulu consultancy, *Sharing the park: Anangu initiatives in Ayers Rock tourism*, revealed the conflict involved with Anangu perceptions of tourism in the park:

> The 'hosts' and 'guests' of this study are perhaps ironically labelled. Anangu hosts want financial benefits from a mass tourism which they did not seek and

with which they do not now wish to interact. Consequently, the more non-Anangu 'guests' who would like to interact with their 'hosts', the less likely such interaction is to occur, except for a very fortunate few. (Mutitjulu Community 1991: 143)

*Discussion*

However, while Anangu involvement is now an integral part of park management, how much authority do Anangu have over decision making in the park? Access within the park is a key policy issue, and one which influences the Mutitjulu community both culturally and socially. Sacred sites within the park are protected by excluding tourist use, and interpretive signs are erected to illustrate the significance of the area. Moreover, tourists often enter areas they have been told to avoid and in the process see places they should not see according to Anangu law. Protests made by visitors to Uluru-Kata Tjuta about being denied access to certain areas have been supported by the Northern Territory's chief minister (Cordell 1993). The visitation rates to the park are also largely beyond the influence of the Anangu, which further inhibits their ability to control, and even influence, activities on their land. The lease agreement itself has created some difficulties within the park, particularly within the Mutitjulu community. Conflict has arisen within the community regarding the distribution of responsibilities and the (mainly financial) benefits arising from the lease. This conflict threatens to disrupt the maintenance of a traditional way of life within the community.

It is argued that such limitations are countered by the many achievements of the joint management arrangement at Uluru-Kata Tjuta, such as the major fauna study carried out between 1987 and 1990. The research project involved CSIRO scientists, senior members of the Anangu community and park staff. The CSIRO researchers and Anangu individuals were employed on the same terms, in recognition of their respective fields of experience. The results of the survey were drawn from both the traditional ecological knowledge of the Anangu, and from the scientific knowledge of the CSIRO researchers. It is believed that the success of the survey was largely attributable to the Anangu ownership of the Uluru-Kata Tjuta land. In some other cases where traditional ecological knowledge has been sought from local indigenous people, the information has been withheld in the belief that the knowledge itself would give the local communities some leverage in their push for land rights. The Anangu ownership and joint management structure of the Uluru-Kata Tjuta National Park were integral to the success of the fauna survey. In addition to the ecological information gathered during the survey, the exercise was successful in strengthening the cooperation between scientists and the holders of traditional ecological knowledge.

The 'Mutitjulu consultancy', which was conducted in 1989–90, addressed issues regarding the implementation of joint management objectives (Woenne-Green *et al.* 1994). The consultancy provided a progress report on the joint management programme and clarified the joint management model by reviewing its implementation from an Aboriginal perspective. It identified a range of issues which were inhibiting the evolution of the joint management process, and brought about a fine-tuning of the Uluru management structure, aimed at 'Aboriginalising' the joint management model. Among the recommendations arising from the consultancy to achieve this was:

- making the culture and experience of the Anangu an integral part of park management;
- formalising power sharing at all levels of park administration;
- establishing routine ways non-Aboriginal staff could gain the advice and participation of Anangu individuals in everyday activities;
- modifying the staff structure to allow easier Aboriginal involvement in park management activities.

Perhaps the most significant factor which has emerged from the continual review of the management model is the turn-around regarding Anangu law within the park. According to Tony Tjamiwa (1993: 7), who has been a member of the board of management since its inception:

> In our park, Aboriginal Law is in the front guiding the way, the board of management doesn't get pushed around, they look after Aboriginal Law . . . There are two lines, the second is the national park, and board of management is number one.

This suggests a radical change in the management style for the park since handback in 1985. Many argue against these sentiments, saying there is a much more balanced approach to management at the park. However, the 1991 plan of management illustrates that Tjukurpa (Aboriginal law) guides the management principles, objectives and strategies designed to achieve the provisions of the lease agreement (Uluru-Kata Tjuta Board of Management & ANPWS 1991).

The shift to the Anangu law as the basis for management has resulted in the unique joint management objectives which have made Uluru-Kata Tjuta National Park a model of joint management worldwide. However, it is important to recognise that the Uluru/Kakadu model is an evolving entity adapting to the issues related to joint management between indigenous peoples and western management techniques within the context of contemporary Australia. As such, it can only provide an example of how joint management may be approached and some of the pitfalls to be avoided. The success of the Uluru/Kakadu model at Uluru-Kata Tjuta is seen to be based on a combination of the power held by the Anangu, the use of Tjukurpa as the basis for management, and the willingness of those involved to implement a successful joint management programme.

## CONCLUSION

Indigenous people have managed land and protected ecosystems for many millennia. This protection was often enforced through complex cultural, spiritual and customary practices and law. We ignore these thousands of years of accumulated knowledge at our own peril.

The rights of indigenous people to self-determination are critical to both conservation and development. Many protected areas, located as they are in remote rural regions, either have indigenous people living in or around them. Therefore the rights of indigenous people represent a particularly important protected area management issue. The results of the colonial and neo-colonial period experienced by indigenous people and the subsequent development of indigenous consciousness mean that no longer can 'first peoples' be seen as the 'other' without rights, legitimate knowledge or legitimate cultural values.

What are emerging are new models of indigenous owned protected areas, managed for conservation as well as achieving the development aspirations of local people.

# 8 Ecotourism

Ecotourism, a non-consumptive use of resources, appears to have the potential to serve both conservation and local development roles well. Tourism is, of course, big business, and hence has the capacity to transfer large exchange, including foreign exchange, to a particular tourism site which can potentially be captured and used for the conservation of natural areas and the development of surrounding communities. Conservation is essentially about values, and any activity that 'captures' or 'uses' the values that people hold for natural resources, and educates about and enhances support for maintaining these values, is likely to be beneficial in a conservation sense. In terms of local development, tourism may provide a vehicle or conduit for translating the values that others hold for a natural area into benefits for those who live in or near it (and may therefore bear costs associated with conserving the resource). The development of this ideal situation is, however, problematic. It assumes that tourism does not impose problems and costs of its own on the natural resources and local people (or at least that if it does, the benefits coming from tourism are greater than the problems it creates).

The major concern of this chapter, then, is whether (and if so, how) tourism can be developed and operated as a local level development activity. Ecological tourism, or ecotourism, is seen by many people as a way in which this may be achieved. In the search for activities and projects that can serve the dual roles of conservation and local development, ecotourism is often thought to be one of the most promising options. In the course of this chapter, we will investigate the concept of ecotourism, the relationship between tourism and local people, the role that tourism can play in conservation and development, and issues and implications of the present global flurry of activity regarding ecotourism. This will be done within the previously discussed social science framework, and will examine social science knowledge and techniques in linking conservation and development through ecotourism.

## WHAT IS ECOTOURISM?

Our perspective on ecotourism may be slightly different to those seen elsewhere, as we are assuming that ecotourism is motivated by the goals of conservation and local development. We recognise that for operators, any tourism must first and foremost be financially viable. We believe, however, that many managers, planners, development workers and local people are interested in ecotourism for its conservation and local benefit, and so our perspective is appropriate, at least for this book.

While ecotourism as a term and a defined set of principles is relatively new, closely related activities (e.g. travel to natural areas) have been practised for many years. The creation of the US national park system late in the nineteenth century was driven in part by recreational interests. Many things appear to be motivating the recent debate and heightened activity regarding ecotourism. The growth in environmental awareness amongst many people, the recognition among conservationists that tourism is one method of capturing conservation values for conservation purposes, the realisation that mass tourism in many cases is destructive of culture and natural resources, and the growing recognition of the general need for ecological sustainability are all contributing in some sense to the development and growth of ecotourism. Before we try to define ecotourism we briefly discuss the tourism experience more generally.

## The Tourism Experience

Tourism of any sort can be described and understood in terms of the experience visitors seek and enjoy. It is important to understand what characteristics and type of experience different areas can (or already do) offer, as well as to understand the type of person interested in such experiences. For places with an existing tourism industry, it is clear that a desirable tourism experience is already available (although this may not be ecotourism). In other cases, there are no existing tourism operations, but a desire to develop tourism is often expressed by managers of protected areas, local or national government, development workers or local people, in the belief that tourism can provide benefits. For these situations it is important that the type of tourism experience available is assessed, and the level of interest among potential tourists investigated. A brief description of some of the principles of tourism management and marketing may serve to clarify the issues involved.

Tourism destinations are often defined according to five categories – attractions, access, accommodation, amenities and activities (Dickman 1989). We will discuss here attraction, although the other issues are also important. Attraction is the element that most people think of in relation to tourism destinations, but attractions can vary considerably. In addition to the well known forms of attraction (such as famous cities, special species of wildlife, historic places, spectacular places, and so on), there are characteristics that can make destinations attractive that may not be immediately perceived. Remoteness, uniqueness, the exotic, the dramatic, different cultures, flora and fauna (not only large mammals or beautiful forests) or an adventurous or challenging setting can all serve to make a place attractive to different types of tourist. While ecotourists will have specific characteristics in common with each other, there can be many different forms of ecotourism. The challenge lies in matching the type of experience available with the type of tourist interested in such an experience, in the most advantageous way for conservation, local people and tourists.

The different types of ecotourist may be differentiated by experience (some looking for a scientific educational experience, some looking for a wildlife viewing experience, and so on), or by other factors such as origin (international or domestic) or length of stay. These factors need to be taken into account in developing a successful ecotourism operation, and the use of social science techniques discussed in this book can contribute to developing this understanding. In some places, it may be possible to offer a number of different tourist experiences, in order to diversify the market (and possibly buffer the local economy against changes in trends).

In addition to dealing with ecotourism at the local level (including accommodation, access and other issues), the international tourism system needs to be taken into account. Marketing, promotions, booking systems, packaging and related issues are all important to the success of tourism. Regardless of how well thought out a local tourist development may be, the linkages into the wider tourism system, and a strategy for dealing with all of the above mentioned issues in a way that will further the goals of local development and conservation, are vitally important. For ecotourism that aims to operate at or from the local level, decisions about the nature of relationships with outside agencies in the tourism industry will have important repercussions for the ownership and control of tourism. There are also some clear advantages involved in making linkages, such as use of existing expertise in the management and booking aspects of tourism, or being able to integrate promotional strategies into existing marketing.

Finally, the nature of tourism as an industry also needs to be understood. Tourism can be highly sensitive to external influences, such as civil strife (even outside the country in which it takes place) or changes in world economic conditions. Substantial concern exists about places which develop tourism to the degree where local people are solely dependent on it for their livelihood, as the success of tourism may change very rapidly according to external factors which cannot be controlled from the local or even national level.

## Definitions of Ecotourism

The definition of 'ecotourism' is problematic. The word has only been in popular usage for a few years, but the popularity of the term suggests that it has struck a chord with many people. The term is generally attributed to Ceballos-Lascurain who apparently coined it in 1983 (Allcock *et al.* 1994), but a confusing and complicating factor is involved in that many forms of tourism (such as outdoor, nature, educational and adventure travel) are now included by many under the umbrella of 'ecotourism', regardless of their characteristics. Filion *et al.* (1994: 236) use the following definition and estimate ecotourism to be worth US$233 billion annually:

> travel to enjoy and appreciate nature.

The Federation of Nature and National Parks of Europe (1993) use 'sustainable' to qualify tourism associated with their parks:

> all forms of tourism development, management and activity, which maintains the environment, social and economic integrity and well-being of natural, built and cultural resources in perpetuity.

The Adventure Travel Society (1994: 2) define ecotourism as:

> environmentally responsible travel to experience the natural areas and culture of a region while promoting conservation and economically contributing to local communities.

And adventure travel as:

> participatory, exciting travel that offers unique challenges to the individual in an outdoor setting.

A good starting point for understanding what people mean when they discuss ecotourism is the prefix 'eco'. To most people, the 'eco' in ecotourism implies ecology. In addition, the connotation that arises from this prefix (partly as a result of the pervasive influence of marketing jargon) is that this form of tourism is ecologically based, ecologically friendly, ecologically sound and ecologically sustainable, or, to use a common phrase, 'environmentally friendly'. Some have, however, pointed out that the 'eco' in ecotourism can also imply economics, as tourism is essentially a commercial activity, and has the potential to contribute to the economic well-being of a locality or region.

Given these interpretations of the prefix, the remainder of the word – tourism – must also be defined. According to one definition in the tourism literature, tourism is basically the 'activity of pleasure travel' (Dickman 1989). Beyond this, tourism also implies the broader tourism system – the commercial and management infrastructure that makes most forms of tourism possible, and the tourism experience. To further confound attempts at definition, ecotourism is used by many synonymously with other terms such as special interest tourism, responsible tourism, alternative tourism, sustainable tourism and adventure travel (Valentine 1993).

Many people, according to Western (1993: 10), see ecotourism as:

environmentally conscious tourism at low volume.

The important point here is the emphasis on volume.

Schematically, we have developed a representation of some different types of tourism to show how they may fit together. In Figure 8.1 all ecotourism lies within the realm of nature tourism, and within sustainable tourism, while some parts of sustainable tourism may occur outside the realm of nature based tourism. This diagram may help to clarify some of the relationships involved between these different terms and the principles they represent.

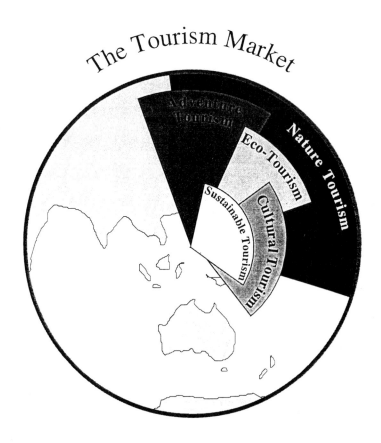

Figure 8.1 The tourism market

## Principles of Ecotourism

Even though the above definitions appear clear and concise, to some extent they do not convey the whole meaning of ecotourism and also introduce new problems (related to the use of phrases like 'ecologically sustainable', for example). Western (1993) and others have seen a need to move beyond restrictive definitions, preferring a set of principles that the term represents. In this way, emphasis is shifted from conforming to a set definition (which may include limits on scale and so on), to a set of ideals that different operations can be both based on and judged against. We have identified a collection of generally accepted principles that can be considered under four headings.

*The setting.* Ecotourism involves natural areas in general, protected areas, and/or places with special biological, ecological, or cultural interest.

*Conservation benefits.* Ecotourism must benefit conservation. Tourism which does not have some form of conservation benefit is not, we would suggest, ecotourism. The benefit can of course be considered as net benefit and include changed community norms through education and consequently political and social priorities and generating income which can be used for management as well as influencing political and social priorities. Of necessity ecotourism must be low impact or at least well managed.

*Benefits to local people.* Ecotourism should generate economic, cultural and social benefits for local people. This may be in the form of increased employment and entrepreneurial opportunities, or equally it may be by way of strengthening specific cultural traits or values. At the very least, ecotourism should have a net benefit on local social and economic development.

*The tourist experience.* Ecotourism should include components of education and interpretation of natural and cultural aspects of a place. Visitors should learn about and develop a respect for the culture of the places they visit, and develop an understanding of nature and natural processes for that place, and through this process, for other places and conservation in general.

Of course, this ideal of a non-invasive, non-consumptive, sustainable, educational and low impact form of tourism may be difficult to achieve. Western (1993: 10) points out that the label 'ecotourist' is being coopted by 'virtually any group remotely connected with nature or culture travel'. Filion *et al.* (1994) consider ecotourism to mean all tourism that takes place in the natural environment. This highlights some problems with the use of such a term. Through coopting the word, groups and agents take advantage of the connotations attached to the word to give their activities legitimacy (deserved or undeserved). Tourism marketing agencies and tour operators have a clear interest in implying to people that their activities are 'environmentally friendly' and benign, as the word connotes. Because of this confusion, it would be more appropriate to assess tourism operators and operations based on the above principles rather than by any claims made or names used.

## ECONOMIC AND CULTURAL IMPACTS OF ECOTOURISM

The remainder of this chapter consists of an investigation of the economic and social/cultural impacts of ecotourism, which will draw on a number of international case studies, followed by a general discussion of these issues.

### Economic Impacts of Ecotourism

*The big picture*

The travel and tourism industry employs 1 in 15 workers worldwide and generated an estimated US$3.5 trillion of output in 1993 (WTTC 1992). High growth rates are forecast, with the figure of 450 million international travellers in 1991 expected to double by the year 2005 (WTO 1992). Barnes *et al.* (1992) quote economic data suggesting that 11% of East African and 13% of Latin American GDP comes from tourism. For nature travel in particular, Filion *et al.* (1994) in their paper on the economics of global ecotourism report that:

- in Kenya/Zimbabwe, 80% of tourists visit primarily for the wildlife; and
- in Latin America, between 50 and 79% of visitors said protected areas constituted an important factor in choosing their travel destination and between 41 and 75% actually visited a protected area.

They go on to claim that nature tourism appears to account for some 40 to 60% of international tourism; and wildlife-related tourism for 20 to 40%. Ceballos-Lascurain (1993) suggests a figure of 7% of all international travel expenditure due to nature tourism. In individual countries the impact of ecotourism can be considerable. Table 8.1 offers data on a selection of countries for which international tourism is a large percentage of the value of exports, giving an indication of growth between 1981 and 1988, as well as magnitude.

Table 8.1 Tourism receipts, growth and export importance for selected countries (after Healy 1992)

| Country | Arrivals 1988 ('000) | Tourism receipts (million US$) | % change in receipts between 1981 and 1988 | Receipts as % of exports |
|---|---|---|---|---|
| Barbados | 451 | 459 | 82 | 265 |
| Seychelles | 77 | 81 | 98 | 260 |
| Dominican Republic | 1116 | 616 | 266 | 69 |
| Samoa | 46 | 10 | 66 | 67 |
| Fiji | 208 | 181 | 35 | 53 |
| Egypt | 1833 | 1784 | 229 | 47 |
| Kenya | 677 | 410 | 86 | 38 |
| Nepal | 265 | 64 | 42 | 33 |
| Belize | 142 | 27 | | 23 |
| Paraguay | 284 | 114 | 25 | 22 |
| Senegal | 256 | 123 | 81 | 20 |
| Thailand | 4231 | 3120 | 260 | 20 |
| Mali | 51 | 37 | 311 | 14 |
| Sierra Leone | 75 | 15 | 50 | 14 |
| Costa Rica | 329 | 165 | 90 | 14 |
| Uruguay | 844 | 190 | −36 | 14 |
| Bolivia | 167 | 65 | 63 | 11 |
| China | 12361 | 2247 | 305 | 5 |
| Brazil | 1743 | 1643 | −8 | 5 |
| Indonesia | 1301 | 1028 | 318 | 5 |

*The local economic impact of tourism*

It is relatively easy to quantify total tourism expenditure, but it is a different matter to track the real impact of tourism in the development process. Healy (1992) discusses impacts in terms of three critical factors: the backward and forward *linkages* between tourism and other sectors; the spatial *location* of tourism activities; and the identity of *beneficiaries*.

The development impacts of an industry depend on the nature of its interactions with suppliers (backward linkages) and its customers (forward linkages). The capacity to stimulate secondary growth and provide opportunities for investment and entrepreneurship depends on these linkages. Tourism, being either an export commodity or, in the case of the domestic sector, a final consumer good (Healy 1992), has few forward linkages – it is not a step in the production of additional activities. It has, however, many backward linkages for food, transportation, construction and so on. In small and/or developing economies, however, many of these linkages extend beyond the country to become imports, with a consequent loss of potential economic activity.

Tourism can be spatially concentrated as activity will occur around particular popular attractions. Many of the economic impacts will therefore be restricted to these areas.

A particular form of tourism in remote locations has sometimes been called 'enclave tourism' (Healy 1992). This occurs where the types and locations of facilities are not oriented towards the local community, and local people cannot afford the services offered. Money generated in these enclaves generally has little effect on the local economy or even on the host country, especially if they are owned by foreign interests. Sometimes these enclaves are deliberately developed in order to insulate local people from the influence of the tourists – for example, to preserve local culture. Given that enclave tourism has few linkages into the local community, and therefore little local development effect, it would be inappropriate to encourage it if local development is considered a goal.

**Plate 8.1** Tourist safari, with elephants at waterhole in Chobe National Park, Botswana. This Zimbabwe based company provides tourists with a safari experience, but there is little return to local communities. (Photo: R. Thwaites)

An important aspect of tourism's economic impact is how benefits and costs are distributed within the community. Local elites may monopolise entrepreneurial opportunities, or businesses may be owned by non-locals (for example, expatriates or urban elites). The distribution of benefits over time can also be of concern. Seasonal tourism may or may not be complementary with other activities, cultural as well as economic, when local people are involved. Most importantly, those who bear the costs (reduced access to resources, for example) often do not collect the benefits. Frequently, without deliberate strategies these problems and questions are not adequately dealt with.

*Property rights*

Tourism depends, to a major extent, on what Healy (1992) refers to as 'background tourism elements' – beaches, forests, rivers, lakes, trails, views and so on. All of these are public goods or partial public goods (see Chapter 3), and this raises a fundamental problem for protecting these resources. The so-called 'free rider' problem occurs where, because of the difficulty of either excluding people from using the resource or paying for the real benefit derived from use, there is no incentive to invest in preserving or managing the resource. For example, the grasslands of Xilingol Biosphere Reserve, Inner Mongolia, provide a magnificent grassland landscape for tourists to visit. However, the cost of maintaining the grasslands, as opposed to converting them to fenced pastures of wheat or planted grasses, is borne by local herders who currently obtain little or no benefit from the tourism. Protected areas are particularly vulnerable to these property rights problems, with grave difficulties involved in capturing even a small proportion of the true economic value of the resource.

Much has been written on how the poor allocation of property rights contributes to environmental degradation (see Young 1992, Chapter 4 for a particularly good analysis of the issue) but in terms of the themes in this book and chapter some points need to be made.

First, a lot more can be done in implementing user fees. How to do this will be looked at later in this chapter. Secondly, two cardinal principles in effective resource right allocation policies are:

1. *Fully specify* both rights and obligations – a fully specified system should make clear who can use, who pays, who can profit, who can degrade and who can control.
2. Each particular use in an area should, as far as possible, be *separated* into an individual bundle with rights and obligations specified for each resource right. For example, property use rights might be allocated to different people or groups over the one geographical area for forestry, grazing, subsistence hunting, water catchment and tourism. The actual owner of the land might be commercial, government or private indi-

vidual, but each of the resource uses, together with obligations, is allocated. Hence, an easement over certain parts of the land is allocated (and perhaps subsequently sold) for tourism activities with certain controls imposed (for example, maximum permitted environmental impact or a requirement that 75% of all employees be local residents).

All this may seem impracticable, but in fact some of the more successful recent integrated conservation development projects which include an ecotourism component are tending to move this way. As part of Zambia's Lupande Development Project (Lewis *et al.* 1990, cited in Lindberg & Hawkins 1993) local safari hunting concessions for South Luanguia National Park are auctioned to tour operators. Assuming the auction is held competitively, this method of selling hunting rights should maximise revenue. Proceeds are then channelled through a fund, with 60% used for wildlife management and 40% given to local chiefs for community development projects. It is reported that poaching of elephants and black rhinos decreased tenfold between 1985 and 1987 after introducing the scheme. As well, the programme generated significant local employment. The CAMPFIRE programme in Zimbabwe (described in detail in Chapter 9) allocates wildlife property rights to community district councils, subject to certain responsibilities including endorsement by community members and implementation of a wildlife management plan. This has allowed councils directly to manage and profit from hunting and viewing tourism and, as a result, pay a dividend to households as well as fund wildlife management and community development activities. At Uluru-Kata Tjuta National Park in Australia (described in detail in Chapter 7) the land is owned by the local Aboriginal community but the tourism property rights are allocated to the Australian Nature Conservation Agency, with responsibilities to return 25% of entrance fees to the owners; to provide employment, training and entrepreneurial opportunities to Aboriginal community members; and to manage tourists in ways sensitive to Aboriginal cultural norms.

Clearly, the way that resource rights are allocated will depend on the local cultural, political and economic situation. But it is also clear that the tradition of government appropriating all resource rights and centrally managing them has not worked to the benefit of either conservation or local development. This is especially the case if they cannot afford to allocate significant resources to fund effective management agencies, implement clear and effective policies or have in place sound local and regional checks and balances in the political system.

## Capturing economic value

Recently there has been much research from a theoretical, policy analysis and practical case implementation point of view, about using tourism fees to capture more adequately the benefits of natural areas for conservation and local development. Lindberg & Huber (1993) have reviewed this issue in an especially communicative way and our short discussion here draws heavily on their presentation.

Before we continue, some points need to be made. The total economic value of natural areas is made up of many components (see Chapter 3 for more discussion) such as consumptive use values, catchment values, bequest values, subsistence values, recreation values and existence values. Many of these components are non-market values (such as the non-use values 'existence' and 'bequest') and hence difficult to estimate and consequently generally not included in decision-making processes such as benefit–cost analysis. They are nevertheless crucial, and should be recognised as important components of value, especially because in many conservation issues these values may be significant but not easily measured, while resource extraction values may be easily measured and can therefore bias management decisions.

Given the fact that, in many countries, the full value that a visitor holds for a place or experience is not captured through the tourism system (usually because fees and costs to the tourist are low compared to what they would be willing to pay), the question becomes one of how the excess value can be captured and put to use in the place where the tourist has the experience.

The first issue is to determine management and revenue objectives (Lindberg & Huber 1993). Are fee revenues to cover visitor management costs, for conservation management (at the site, in the region or country), to contribute to local development programmes or all of these? Are costs to be set so as to subsidise recreation for local, regional or domestic visitors? Many countries have set differential fees for foreign tourists and domestic tourists, in order to address access and equity problems that can arise from pricing according to the willingness to pay of affluent western tourists. Are fees going to be used to regulate seasonal demand or disperse visitors by varying fees at different times or places? Fees may be kept low to attract tourism to increase business activity in a region, or alternatively raised to 'market' prices to encourage private development of alternative ecotourism facilities. Notwithstanding all these possible objectives, it would seem important at least to set tourism fees or charges for international tourists to cover the costs of providing the ecotourism opportunity, including operating costs, capital costs and especially indirect costs such as ecological damage and negative impact on local communities.

How then do we set tourism fees and charges? As mentioned in the previous section on property rights, many of the natural resources that form the so-called 'background tourism elements' are public/non-rival, non-excludable goods and, as such, it becomes difficult to set a 'price'. Common theory suggests that prices should be determined by the willingness to pay of the visitor.

Lindberg & Huber (1993) suggest four techniques for estimating fees to meet objectives.

*Market evaluation*  This technique is based on estimating the demand and cost involved for similar attractions. Finding a similar attraction/destination package may be difficult and adjustments need to be made for any significant variations when comparing the place in question to others. Issues which may vary and so affect demand include: competing attractions (Nepal's Mount Everest is unique, being the highest, and so there is no directly comparable attraction); attractions (for example, the presence of large carnivores or exotic cultures); complementary attractions; political and economic stability; the visitor experience (the more enjoyable the overall experience, the higher the fees that can be charged). Other issues affecting demand include the type of tourist who is likely to want a certain type of visitor experience.

There is surprisingly little reliable demand information for ecotourism, so there needs to be a fair degree of intuitive judgement when estimating the importance of each of the demand factors. Nevertheless significant experience is becoming available in the tourism literature of different tourism charges that have proved suitable, both in raising revenue and maintaining visitation at appropriate levels.

*Survey of tourist demand*  In this method, tourists are asked what they would be prepared to pay for either the existing experience or, more commonly, changed or improved facilities – in short, they are asked to estimate their own demand. Opinions, via visitor surveys, can be obtained on fees and charges for both resident and overseas visitors (for information on administering surveys, see Chapter 4). Some problems can be encountered in using surveys for these type of demand evaluations, and concerns are expressed about underestimation of demand and other problems; but, if survey techniques are applied correctly, valid information should be readily obtained. Contingent valuation surveys are a more rigorous means of obtaining valid and reliable economic figures, and offer another option for demand evaluation.

*Demand curve analysis*  This is the most accurate strategy to set fees to maximise revenue or, if converted to marginal revenue curves and compared with marginal cost curves, can be used to set profit-maximising fees, although estimation of demand curves can be difficult. Where demand curves have been analysed, interesting data sometimes emerges. Lindberg & Johnson (1994) report that for foreign tourists visiting Belize national parks the revenue maximising fee would be US$367 – some two orders of magnitude higher than park entrance fees levied or contemplated. Trying to charge this as entry fees could cause considerable consumer resistance, but collecting money via indirect means such as airfare levies, hotel taxes and so on may be possible; although it must be coordinated by government rather than an individual protected area.

*Market-based reactive management and auctions*   Here, fees are charged in reaction to consumer demand. A fee is set as a result of estimating market demand or a tourist survey, and then adjusted depending on demand – upwards if visitors increase or conversely downward if demand falls. Auctioning of permits for hunting or trekking is a variation on this method. Government-run ecotourism operations, indeed government-run operations of any sort, rarely have the flexibility or local autonomy to implement market-reactive approaches – which is a good reason for tourism resource rights to be allocated to a locally based institution.

*Types of fees*
Once fee objectives have been established and levels estimated, managers need to decide what to charge fees for and how they should be collected. Lindberg & Huber (1993) suggest the following alternatives.

*Direct on-site collection*

- entrance fees – to enter park, for example
- admission fees – for a specific facility such as a cultural centre, for example
- use fees – for ranger-guided services, or camping places, for example
- permit – for hunting, fishing and so on
- concessions – a leasing agreement with an entrepreneur
- sales – souvenirs, for example
- donations – for conservation of an endangered species, for example

On-site collection has the advantage of directly correlating service with charges (Lindberg & Huber 1993). In addition, collecting money at the local level may make it easier to respend at the local level, on conservation management and local development. One disadvantage is that there may be consumer resistance to high fees if they are applied on-site, especially if fees are low in comparable areas.

*Collection through tour operators* This method can help in improving linkages to local business and the wider tourism industry (Lindberg & Huber 1993). Charging indirectly, by levying tour operators a fee per person (or per trip), can often reduce administrative costs and 'hide' the charge from the tourist in a larger package.

*Indirectly, through other sectors of the tourism industry*

- transportation taxes
- hotel taxes

The collection of fees using these methods can provide benefits similar to those received through tour operator fee systems (Lindberg & Huber 1993). There may be difficulties with the tourism industry accepting such fees unless there is a clear linkage between the fees charged and the use of the revenue collected.

*Some examples* An international survey (Gionjo *et al.* 1994) shows the combination of sources from which protected areas receive their revenue. Roughly 50% of protected areas charge entrance fees. However, entrance fees generally comprise: a very small percentage of total tourism expenditure (when airfares, accommodation and other costs are taken into account); a relatively modest proportion of protected area management cost (figures for which are surprisingly rare); and in most cases probably represent a tiny fraction of the total economic value of these protected areas.

Lindberg & Enriquez (1994) describe a number of examples of tourism charges which we summarise here:

- In Latin America fees are generally low or non-existent. Costa Rica's park entrance fees are quoted at US$1.40.
- The Galapagos National park, Equador, has recently increased fees for foreign tourists from US$40 to US$80, generating approximately US$3 million annually directly for Ecuador, of which 11% goes to manage the park. The fee for Ecuadorians is US$6.
- Many protected areas in eastern and southern Africa charge entrance fees of US$10 to US$20. Kenya increased entrance fees for foreigners from US$10 to US$20 in December 1993 and plans to cover system-wide wildlife management costs through tourism-related revenues. It is

anticipated that the Kenya Wildlife Service will earn $53.7 million in 1995 (Kenya Wildlife Service, 1990, cited in Lindberg & Enriquez 1994).

- In Australia the Great Barrier Reef Marine Park Authority introduced a US$0.70 per person per day environment charge to be levied on all tour operators – the money to be used for reef management, research and education.

Table 8.2  Protected area revenue sources – % of protected areas citing each source (after Gionjo *et al.* 1994)

| Revenue source | US/Canada | Eastern Europe | Asia | Africa | South Africa | Australia | Latin America |
|---|---|---|---|---|---|---|---|
| Entrance fees | 58 | 40 | 50 | 68 | 62 | 30 | 42 |
| Concessions | 25 | 20 | 19 | 16 | 6 | 20 | 7 |
| Licences | 9 | 13 | 7 | 10 | 6 | 10 | 5 |
| Donations | 18 | 7 | 23 | 23 | 6 | 10 | 12 |
| Foundations | 9 | 27 | 10 | 10 | 0 | 0 | 12 |
| Admin. agency | 82 | 94 | 87 | 80 | 73 | 73 | 71 |
| Other | 13 | 33 | 7 | 20 | 23 | 50 | 15 |

## Cultural Effects of Tourism

Negative effects on local culture are often attributed to tourism. In a general sense, many of these problems are related to impacts associated with any change, with tourism being the agent of change. As discussed in Chapter 3, in all social systems there is a set of cultural norms and values that regulate social institutions and human behaviour in that society. Culture consists of shared thoughts, beliefs and values (such as religious orientation, philosophical systems, conceptions of law and justice), various behavioural components (institutions) and material expressions (artifacts, houses, public buildings and so on). Culture provides the 'soul' of a community with its own cultural sensitivities and aesthetic expressions. Where culture is made to serve tourism it is simultaneously being transformed into a market-oriented commodity – the so-called commodification of culture (Dahlan 1990). Dahlan further classifies this process as first the transformation of cultural legacy, historic heritage and national endowments into determinate commodities for a tourism market – commodification – and then the creation and control of tourism markets for these commodities – commoditisation of culture.

Three elements interact in this commodification process: politics, culture and ecology. If a group's culture and ecology (their natural endowment) have been commodified, without the group perceiving that it has contributed to their needs or has led to a degeneration of the symbolic order of the group, there may be significant political protest and opposition to tourism. Vulgarisation of culture, common to much of tourism, can also generate opposition and resistance from host communities. Dahlan (1990), reviewing a number of international case studies, discusses two patterns of tourism that cause significant cultural problems. The first is where the business culture of tourism has not extended widely enough to incorporate the effective participation of local groups. Tourism is perceived as either international business or part of the socioeconomic development of the host country, and there is no niche for local groups in the overall development process. The second pattern is when tourism destroys the basis of the host society's moral and symbolic order. This also leads to the vulgarisation of culture and can contribute to the destruction of that culture.

Perhaps at the heart of the problem is the failure of tourism (as a market operation) readily to consider issues other than utilitarian financial ones. While the economic imperatives of tourism are crucial, it must also deliver as an agent of intercultural exchange providing a greater understanding of different historical, cultural and ecological experiences that have shaped different societies today. When we deal with culture we are dealing with symbols that communicate the soul of a host community. Thus, when tourism interacts with culture, we must be deeply concerned with avoiding cultural misconceptions and violations that may in the end destroy or devalue the tourism industry for that area.

The accusation is often levelled that tourism is a form of neoimperialism or neocolonialism. Those who make these claims argue that tourism is another form of exploitation imposed by western nations on those in developing countries. Developments such as foreign-owned luxury hotels and resorts, or where local people are no better off than prior to development, tend to support these claims. These accusations in part are what have led to the surge of interest in less exploitive forms of tourism such as ecotourism.

A brief outline of some of the ways that social and cultural problems can arise may serve to highlight areas that require special attention:

- Different cultural norms and assumptions (about child labour, the role and status of women, religion, alcohol, and so on) may be shocking or offensive to tourists or to locals.
- The local economy and even family economies may be disturbed through the introduction of tourists and tourism earnings – even small amounts of foreign money may have the spending power to make some local people very rich, or to allow a child to earn more than a parent in some cases (Dickman 1989). Diversion of labour from essential activity such as food production to more lucrative tourism work may also be a problem (Dickman 1989).

- Prices of goods and services in local areas may rise to meet tourist demand to the point where local people can no longer afford them.
- Displays of wealth and the attraction of the consumerism displayed by tourists (even at the level of personal possessions such as clothing) may be disruptive for local people who aspire to similar conditions.
- Restriction of access to resources, diversion of resources and different rules for tourists may (justifiably) cause resentment among local people.
- Presence of tourists and their desire for souvenirs and artifacts may stimulate trade in culturally important items, and even the mass production of replicas may 'devalue' the original items (Hudman, cited in Dickman 1989).
- In many cases, tourists (especially if they expect 'cultural' experiences) feel they should be privy to (or even able to participate in) cultural rituals, occasions or activities (such as dances). This can be a problem where local people feel they are obliged (in order to keep tourism happening or to make more money) to perform cultural acts as a form of entertainment, thereby destroying or changing the meaning of them. For example, the commercialisation of fire walking in Fiji is described by Rajotte & Bigay (1981).

The main mechanisms for avoiding difficulties in ecotourism may include: fully briefing tourists prior to their visit as to culturally acceptable behaviours (local people may need to inform tourism managers of what they require); ensuring tourists have a good understanding of their host culture, beyond a simplistic and idealised notion (through education as part of ecotourism); designing tourism facilities and systems in such a way that local people are comfortable with their role in tourism (such as developing accommodation situations where tourists stay on the terms of the local people, and local people are not relegated to menial work); planning tourism carefully where possible to avoid negative impacts; developing structures to ensure the localised economic effects of tourism are not destructive to the local economy; and ensuring that local people are aware of the different possibilities that tourism may bring.

### Desert Tracks in Outback Australia

One example of ecotourism operated and managed by local people (in this case indigenous people) is 'Desert Tracks', an operation that offers tourists a desert living experience with Aboriginal people in outback Australia. The Pitjanjatjara Land Council, an organisation that manages land and affairs of the Pitjanjatjara people, has developed the tourism operation in an effort to control what tourists see and learn on their land.

The tourism experience offered by Desert Tracks involves small groups (around 12 people) who are transported from Alice Springs to the Mann Ranges to live with Aboriginal people, sharing in work tasks and food, and sleeping out using sleeping bags and 'swags' (canvas protective covers). The tourists are guided by Pitjanjatjara speaking people, who fully instruct them in their visit, including advising them on correct behaviour.

The Aboriginal people who host the tourists offer experiences such as gathering bush food and the chance to be involved in ceremonial dance events. The fact that the local people have established tourism on their terms, and under their control, demonstrates the potential for this to occur. The small size of tourist groups, while limiting revenue potential, allows the experience offered to be optimised and enhances the quality of the visitor experience. Premium prices can be charged for high quality or unique experiences, providing an alternative to large numbers of tourists paying smaller amounts.

Source: Kovall (1994)

## CASE STUDIES

At this stage we present four case studies: ecotourism in Nepal; ecotourism at Ranomafana National Park in Madagascar; the social impact of tourism in Sagarmatha National Park in Nepal; and ecotourism in Belize. These case studies provide us with examples of how the social sciences can be used to analyse ecotourism development initiatives. The case studies cover four different types of analysis – a macroeconomic analysis of ecotourism for an entire country, a microeconomic analysis of ecotourism in a single protected area, an analysis of the social impact of ecotourism, and an integrated analysis. The case studies will also serve to show some of the strengths, weaknesses, successes and failures of these operations or projects, and how people in different parts of the world and different situations have addressed challenges.

### A Macroeconomic Analysis of Ecotourism in Nepal*

Michael Wells (1993) conducted an economic analysis of the impact of tourism in Nepal. Nepal has seen a remarkable increase in foreign visitation – from 6000 in 1962 to around 260 000 in 1988. Most of these visitors have travelled to see the country's capital city, Kathmandu, or the spectacular protected areas, chiefly those in the Himalayan mountains. One circumstance that facilitated Wells' analysis was that until recently travel out-

---

* This case study is based on the economic analysis of the impact of tourism in Nepal carried out by Wells (1993).

side the capital has been limited to a small number of areas, all of which include protected areas, making it easier to determine the proportion of tourism that is attributable to protected area visits.

The basic conclusion of Wells' analysis is that, although a substantial amount of money is earned through tourism that can be directly attributed to the protected area system in Nepal, this amount does not reflect the amount that people would be willing to pay to visit, and much of the money earned from tourism leaves the country. In addition, very little of this money is actually captured by the protected area system of permits and fees. This has implications for the funding of protected area management – it could be argued that policy and decision makers are less likely to allocate resources to protected areas when they do not directly make large amounts of money. The official estimates of the total amount of money received from tourism in 1987/88 was $76 million, although unofficial estimates (which include black market exchange activity) were around $113 million for the same years. This places tourism as the leading foreign exchange earner for Nepal, and around 11 000 people were directly employed in the tourism sector in 1988.

*Economic benefits of tourism to protected areas*
Wells estimates that around 20% of the total visitation to Nepal is directed towards one or more of the protected areas in the country. Separating out the economic effects of protected area tourism from total tourism receipts can be very difficult. Some estimates have been derived indicating that around $26.8 million of 1988 tourist expenditures were directly related to protected areas. It is also estimated, however, that two-thirds of this amount left the country to pay for imports of products not locally available. Therefore the best estimate that Wells established was that $8 million (in 1990/91 figures) attributable to protected area visitation was retained in the country. This figure is subject to certain inaccuracies – visitors may have been motivated to come to Nepal for reasons other than visiting the protected areas, and the figure does not take into account multiplier effects of tourism monies in the domestic economy or expenditure on international air travel.

*Money directly captured in the protected area system*
In addition to general tourism expenditure, money is obtained from tourists from protected area entry, trekking and mountaineering permits, and concession fees for operations in the protected areas. The total monies from these sources directly related to the protected areas were around $1 million in 1990/91. The general consensus regarding these fees is that they are well below the amounts that could be charged. This indicates that the consumer

surplus obtained by visitors is high, or in other words the economic value that Nepal could have benefited from is being lost, instead of being retained as actual revenue in some form.

*The costs of the protected areas*

It is clear that there are substantial existing and potential benefits from the protected area system. What, then, are the costs of keeping the system? Wells estimates the management costs of the areas, and also discusses the indirect costs and opportunity costs involved. The management budget for the total protected area system amounted to $4.6 million. Much of this money is spent on law enforcement via the Royal Nepal Army, the remainder amounting to less than $1 million for 1990/91. This budget is regarded as insufficient for addressing many of the problems that are facing the protected areas in Nepal (although there is some debate about the magnitude and causes of these problems). Other costs include unmeasured indirect costs related to the damage that may be engendered by protected areas. In Nepal's case, damage caused by protected wildlife, such as tigers, bears, rhinos and leopards, is most common and includes death and injury, loss of livestock, and crop damage. Opportunity costs – the costs foregone because the areas cannot be used for other activities – include opportunities for hunting, forest product collection and livestock grazing. These costs have not been measured, although they may be of considerable importance – especially as they are usually borne by local people who receive no compensation. Other opportunity costs would include the best alternative land use such as hydroelectric power generation or forestry. Although unmeasured, opportunity costs of this sort are likely to be low – Wells concludes that the most economically attractive land use for the protected areas is tourism.

## An Analysis of Ecotourism Economics at Ranomafana National Park, Madagascar*

Ranomafana National Park is located in an area of rainforest in southeastern Madagascar surrounded by predominantly poor rural communities. Tourism is the third largest industry in the country, which is characterised by vast biological diversity. Ranomafana National Park has high conservation value due to the relatively intact rainforest habitat and the fauna it supports, including 12 species of lemur. The park is adjacent to the town of Ranomafana, which is a long established tourist destination.

---

\* Community development and conservation in Ranomafana National Park has been reported on by various people and this analysis draws on several of these, but especially the reports by Peters (1994) and Wright (1993).

# ECOTOURISM

In 1987, the Ranomafana National Park Project was instigated, as a result of the coordinated effort of the Malagasy Ministry of Water and Forest, the Ministry of Higher Education and American university partners (Duke University, North Carolina University, Yale University, Washington University and Missouri Botanical Gardens). One of the first exercises in the national park was the evaluation of development problems in the villages surrounding the park. These interviews made clear to the project team and government officials that a plan for the national park must incorporate the needs of the local people. A large number of international development, conservation and funding organisations helped support biodiversity research, training of local people in science and protected area management, sustainable development/production systems, socioeconomic and health studies, education, park management research and ecotourism analysis. Over 150 researchers are involved in the overall project. The project has a very substantial budget, with at least $3.2 million available. It is therefore by no means a small-scale exercise, nor is it one that could be emulated without significant access to international conservation funding agencies, links that would have been unavailable without the university partners.

Peters (1994: 3) states that the basic rationale for the ecotourism component of the project was 'to stimulate the local economy and generate revenue for the sustained operation/maintenance of the park'. Similarly, Wright (1993: 13) feels that it is 'clear that in the long run, the park has to generate income on its own and help pay for itself'. Based on these broad objectives, a one-month feasibility study was designed and conducted, examining the socioeconomic conditions of the local people and the constraints and opportunities that would affect ecotourism development. A social impact study was subsequently conducted, which found that the existing distribution of tourist monies was uneven in the area, in terms of the amount that is respent locally and the different groups in local society that benefit. In total, around half of the revenue ascribed to tourism is respent locally. The unequal distribution between social groups is more striking. The majority Tanala ethnic group owned and operated only 2% of the local businesses related to tourism. The Betsilio ethnic group, which is local in origin, is involved in 50% of the businesses, but the Merina group, which is in a minority and originates from the capital city area, operated 45% of the businesses. The study also showed that only around 65 local people benefited from tourism, representing only 0.002% of the people surrounding the park. The logging that was conducted prior to the park's declaration employed people numbering in the hundreds, one loss which is directly attributable to the park. Clearly, mechanisms must be developed for wider and more equitable distribution among local communities of the benefits from tourism.

In 1992 an ecotourism study for all of Madagascar's protected areas was commissioned. The main recommendation of this study was that a revenue-sharing plan be introduced whereby 50% of protected area gate receipts would go to the National Association for the Management of Protected Areas, and 50% would go to a locally organised and controlled revenue management committee for each protected area. The operation of the plan was dependent on a mechanism at the local level for management of funds and allocation of decision making, and protection from corrupt practice. The provisions of the plan were applied to Ranomafana National Park through a general assembly composed of representatives from the various villages around the park. This assembly was to oversee the spending of the funds from park entrance revenue, as well as serving as a forum for the discussion of issues relating to the park and local people. In addition, a management committee was to be formed to choose from different village development proposals and allocate the funds accordingly.

In addition to the revenue-sharing plan established by the National Association for the Management of Protected Areas, the project developed a number of proposals for dealing with ecotourism development issues. The project initiated a training programme for 20 local people to become tourist guides and encouraged the upgrading of the existing hotel and the establishment of tourist bungalows (Wright 1993).

*Discussion*

The main point emerging from this case study is that social analysis is necessary, first to establish the existing situation, and secondly to plan projects that will have the desired results and hopefully no unexpected consequences. The inequity in the distribution of revenues identified in the case study is a critical element in the local people/park relationship, especially as the people benefiting least appear to be having the most impact on the park. It remains to be seen whether the institutions put in place to redress the imbalance will ameliorate this problem. A deeper social understanding may reveal issues of disempowerment that might prevent the majority ethnic group from taking advantage of the new revenue-sharing mechanisms. An attempt to understand local social issues in more depth may also lead to more varied and creative initiatives for equitable distribution of benefits from tourism. To some degree, reservations could be expressed about the nature of the institutions and mechanisms developed by the project for revenue distribution. The nature of the assembly and management committee, with the process of application for project funding, may serve to disadvantage further local people who do not have the skills to work well in such a system – perpetuating the dominance of a certain group of people. The involvement of local people in conceiving and developing plans for tourism may also have been neglected. The vast support at many levels offered by the project may be difficult to withdraw; these issues could be better addressed with a good understanding of the local social situation. Peters (1994: 4) recognises that these issues are of concern and hopes that 'in the long run this interactive planning process may lead to real power sharing and co-management of park resources by all three partners' – the National Association for the Management of Protected Areas, the park managers and local people.

## Cultural Impact of Tourism in Sagarmatha National Park, Nepal*

Sagarmatha (Everest) National Park in Nepal is famous for both Mount Everest and the culture of the local Sherpa people. The culture of the local people comes from Tibetan origins, and their traditional lifestyle included agropastoralism and trade with Tibet, and is strongly influenced by Buddhism. Within the main area of the National Park there are eight Sherpa villages and over 100 seasonally used settlements. There are around 3000 people living in these villages. The actual villages and seasonal settlements are not controlled directly by the park management, although Sherpa grazing and forest lands are.

The area has seen a dramatic increase in tourists, nearly all of whom come to trek in the Everest region or are mountaineers. As a result of the influx of tourists and money, the local Sherpa economy has moved rapidly from a subsistence, agropastoralist basis to a cash basis dependent on tourism. According to Robinson (1994), the local benefits have been substantial and the standard of living of Sherpa people is now at least as high as that of any of Nepal's ethnic minorities.

Robinson highlights the influx of Sherpa people into the trekking industry. In 1989, 26 of 56 larger trekking agencies were owned by Sherpas; 92% of the urban population of Sherpas gain income from the tourism industry, while 80% of the Sherpas who remain in the villages of the park do so. These people are involved in running teashops and lodges or in providing food, equipment and souvenirs. Jobs at all skill levels in the local tourism system are filled by Sherpa people, and some low skill positions provide more remuneration than office workers earn in the capital. The seasonal nature of the industry means that many of the local people revert to subsistence activities in the off season. For Robinson, this indicates that traditional, economic and cultural ties to their customary life remain intact.

However, some villages have not been the beneficiaries of tourism as they are away from the main trekking routes. The increased market prices for food and other goods have also disadvantaged some. In addition, the local economy is entirely based on tourist arrivals, which can be influenced dramatically by political and other events.

---

* Tourism in Sagarmatha National Park and Nepal in general is well documented in the literature. This case study draws mainly on Robinson (1994).

Robinson (1994) chronicles the substantial social changes that have occurred in conjunction with the growth of tourism and the involvement of local people. Local occupations, both traditional or more skilled positions such as carpenter or school teacher, are not as attractive or lucrative to young people as trekking work. The short-term financial rewards of working also discourage young people from moving into higher education.

The local basis of social status has also changed from land and animal ownership to income, employment position and number of pack animals. Families with high status in terms of trekking, as well as becoming more affluent than others, also become more influential, holding even more status than local elders or religious leaders. Time spent away from home is another change imposed by the nature of trekking employment, affecting political institutions in some cases, leading to fragmented interests and a lack of consensus. Demographic changes, caused in part by the move to a cash economy as well as by family planning and by high mortality in young males through mountaineering accidents, are also in evidence.

The degree of 'westernisation' of Sherpa culture is more difficult to assess, according to Robinson (1994). Association with tourism has led to the admiration of western culture by Sherpas. Traditional clothing is less commonly worn and consumerism is more prevalent. The giving of gifts, establishment of business partnerships and travel to the homes of visitors have also changed the Sherpa culture, with many having travelled abroad.

However, despite these changes, Robinson maintains that the fundamental elements of Sherpa culture appear intact and strong. Social research indicates that the core values of Buddhism, a central element of Sherpa culture, still appear to be strongly held, and interest and involvement in rituals and festivals remain important. The results of studies show that Sherpas think of themselves 'primarily and uncompromisingly as Sherpa' (Fisher, cited in Robinson 1994), and that they now value some of their traditions even more than before the start of tourism.

What characteristics and reasons are responsible for this? The nature of the local culture, the types of tourist visiting, the nature of the resource and the local economies have all been important elements in these changes. According to Robinson, one reason that tourism has not had significant negative social consequences is probably the gradual and late (historically speaking) introduction of tourism to the area; another is that the type of tourist coming to the area has characteristics that facilitate beneficial interaction with the local people, and the tourist experience forces close interaction and dependence on local skills and knowledge; while the isolation of the region (meaning that local employment and supply of tourism services are more efficient than importing from outside the area) may also have played a role.

The image the visitors have of Sherpas as 'egalitarian, peaceful, industrious, independent and compassionate' (Robinson 1994) may be an image that they aspire to themselves, and therefore the self-image and view that Sherpas have of their culture may be reinforced and validated by expressions of respect and admiration. The nature of the interaction also appears to promote this, as the personal interaction over relatively long periods of time that is necessitated by the tourism experience on offer gives visitors a feeling of gratitude which also helps to reinforce Sherpas' image of self and their culture as valuable (Robinson 1994).

**Plate 8.2** Sherpa women in the Himalayas in Sagarmatha National Park. Sherpas have benefited economically through establishment of trekking companies. (Photo: J. Bauer)

Much of the research on cultural impacts of tourism highlights its problematic nature. In many ways tourism can be a double-edged sword, with the potential for wealth generation subsumed by the costs in terms of lost culture, dependence on outsiders and, at least potentially, the entrenchment of existing local inequalities as economic benefits fail to trickle down. For Robinson, the negative impacts of tourism have been at least partly offset by Sherpas being able to benefit at village level as well as institutionally through the establishment of trekking companies, and very importantly by the particular nature of ecotourism in Sagarmatha National Park.

### An Integrated Analysis of the Economics of Ecotourism in Belize*

Belize is a small Central American country, located on the Caribbean coast bordering Mexico and Guatemala, with a population of only 200 000. An integrated analysis of ecotourism in Belize was carried out by Lindberg & Enriquez (1994) for WWF. There were three case study sites examined in the integrated study – Hol Chan Marine Reserve, Cockscomb Basin Wildlife Sanctuary and the Manatee Special Development Area. Hol Chan Marine Reserve protects part of the barrier reef off the coast, the world's second largest coral system, with significant conservation, scientific and tourism importance. The Cockscomb Basin Wildlife Sanctuary is an area of regrowth forest established to protect the ecosystem, watershed and the habitat of jaguars (*Panthera onca*) while promoting and facilitating nature tourism. The Manatee Special Development Area is reserved to provide a zoning process that allows for protection of water, flora and fauna, as well as for low density housing, small-scale agriculture and commercial use. In all three reserves there are opportunities for sustainable use of the resources for the economic benefit of local communities.

The integrated study of ecotourism assessed the economic (financial) impact of ecotourism spending on two of the three protected areas in the case study; the Manatee Special Development Area has no management structure and therefore does not collect revenues or spend money. The focus of the work is on foreign tourism – visitation by Belizians is included as 'traditional management activity' for its environmental education functions. As the scope of the study was to assess the financial impact of ecotourism, the economic and social impacts were not considered. Revenues (or benefits) attributable to ecotourism in the case study assessment included international donor support, bank fees, on-site donations, merchandise sales and entrance fees. Other benefits include educational and scientific opportunities, and the social values related to conservation and biodiversity.

The case study also examined the external, opportunity and direct costs of tourism within the reserves. Ecological costs of ecotourism involve damage to the resource or consumption of resources caused by tourism. Some erosion at Cockscomb has occurred as a result of tourism, but management actions have been able to minimise these (at a cost). Hol Chan suffers reef damage from tourist activity and hotel discharges. Social costs are negative impacts that tourism may cause to local communities. Inflation has been one of the main social costs of tourism, as well as a possible increase in crime and congestion costs which may be reducing satisfaction levels for tourists, especially at Hol Chan. The direct costs of tourism were calculated in terms

---

* This case study is based on reports by Glaser & Marcus (1994) and Lindberg & Enriquez (1994).

of wages, maintenance, facilities, expenses, infrastructure provision and interpretation.

The authors of the case study did not conduct a full financial analysis, as protected areas are new in Belize and no long-term data are available, and there are no explicit financial objectives for protected areas in Belize. However, it was obvious that ecotourism activities at Cockscomb were at levels which did not allow for full cost recovery. At Hol Chan, revenue from ecotourism activities were just paying for the costs of ecotourism.

The researchers examined a number of options for increasing tourism-related revenues and discussed ways of gaining revenue through fees. The vast majority of visitors surveyed in Hol Chan in one study indicated that they would be willing to pay more than the existing BZ$3 fee. The basic conclusion of the discussion is that fees could be substantially increased without affecting visitation. The authors established that fees of an amount unlikely to change visitation substantially could be imposed for Cockscomb and for Hol Chan to cover ecotourism costs and generate revenue for management actions. The researchers also analysed the national effects of tourism, which revealed that it is a multimillion dollar industry important to the country as a whole.

Resident surveys in villages and towns near the three case study sites under discussion were also carried out in order to study the effects of tourism on local communities. The surveys found that relatively little risk (in terms of investment or of sacrifice of other employment) has been necessary to obtain existing tourism-generated benefits. In the villages and towns studied, the proportion of households obtaining benefits from tourism ranged from 67% to 0%, and many of those households who gained no economic benefit from tourism still expressed that they had personally benefited from tourism development in the community. Ownership of major components of the tourism industry, such as hotels, was identified as a concern, as many of these are foreign owned. The survey also investigated whether people who traditionally used resources from the protected areas have benefited from tourism. The results show that many of the households that have suffered from the establishment of the protected areas have gained benefits from tourism, but that many have suffered costs (due to resource restrictions, damage from wildlife, or inflation due to tourism) and have not received direct tourism benefits.

The final component of the integrated study involved an investigation of the attitudes of local people to conservation, tourism and resource use. Issues identified included: reduced access to resources; damage to crops or wildlife; costs of tourism development; employment opportunities as staff; economic benefits from tourism; productive and other benefits; the attractiveness of the resource as opposed to other alternatives; and the likelihood of punishment for illegal resource use. In addition to these factors, the authors point out that a variety of social and cultural considerations will also have an effect on attitudes, and discuss the fact that attitudes and actions of people near protected areas depend on costs and benefits to them from that protected area. The authors concluded that while economic benefits were accruing to the local communities, ecotourism could play a major role in sustainable resource use within the reserves.

## CONCLUSION

The above discussions and case studies provide us with examples of how the social sciences can be used to analyse ecotourism and development initiatives in two ways – first, analysis can be made in the early stages or prior to an ecotourism operation beginning, to allow for strategic planning or policy development and secondly, analysis can inform policy and planning changes for existing ecotourism enterprises. Many ecotourism activities already exist but urgently require analysis of their social and economic context and consequences, while many others are starting up, claiming the title ecotourism (and often claiming to benefit or involve local people). Both the existing and new tourism operations and projects can benefit from social and economic analysis.

**Issues Relating to Local Development**
Brandon (1993) suggests three basic reasons for ecotourism not having led to ecodevelopment: absence of political will by governments to ensure integration of ecological principles with economic development; the fact that tourism is often promoted by large-scale interests from outside the local area with the resultant high leakage of benefits; and a lack of integration of local needs and preferences into the planning process. She then goes on to describe a number of specific issues critical to ensuring community participation in ecotourism. Using her headings we conclude this chapter where we began our book: how can we link development to conservation through protected areas (and ecotourism)?

*Empowerment of local people*
As described in Chapter 1, local participation means empowering people to make meaningful decisions about their own lives rather than being passive recipients of others' actions and decisions. Local people can then become an important driving force in the development process. Brandon compares the 'participatory approach' which empowers local people to the 'beneficiary approach' in which local people receive benefits but are not central to the development process. She asserts that many ecotourism projects fail because they use the beneficiary approach. An example would be where an eco-

tourism project is designed which potentially creates substantial local economic activity by way of jobs as guards and guides, food and craft sales and accommodation opportunities. But, because local people were not part of the decision-making process, problems are encountered. Labour is not available at high tourism demand times as it conflicts with harvest; a particular ethnic group monopolises business opportunities, causing major conflict in the community; or poaching of wildlife increases as disaffected groups seek their own rewards.

*Participation in the project cycle*
As highlighted throughout this book, participation is central to linking development and biodiversity conservation. We have discussed above the necessity for participation at all stages of ecotourism developments, including pre-project planning, project development, project implementation and project evaluation. Importantly, local people need to participate in the information-gathering or research phases of any project.

*Creating stakeholders*
Crucial to any participatory process is ensuring that all those involved have a sense of ownership of the project. One important way is for local people to become stakeholders in the nature tourism activity, both individually and as institutions. The earlier discussion on property rights is relevant here.

*Linking benefits to conservation*
The linkage of benefits that accrue to local people with the ecotourism activity needs to be explicit in the objectives and strategies of the project. The most effective way of achieving this is to ensure that as many in the community as possible directly (and obviously) benefit from tourism, and that its continuance is seen to depend on conserving the resource. Hence the following are essential to the success of the venture: hiring (and training) many part-time guides rather than employing a few full time; direct payment of a percentage of tourism fees and charges to community groups and householders; ensuring community and school education programmes are developed; and emphasising the linkage between conserving the resources and overall community and social benefit.

*Distributing benefits*
As emphasised in the previous paragraph, ensuring that as many as possible in the local community benefit is important to conservation as well as social equity. Business, which is primarily driven by the need to make profits, will make decisions based on efficiency rather than equity considerations.

Governments, on the other hand, are often susceptible to pressure from power elites. Distributing benefits widely is thus not easy. Brandon (1993) lists some practical questions to ask:

- Is it better to have a hostel and restaurant run by the community with revenue sharing or for individuals to establish lodges?
- Is there local capacity to run business efficiently? If not, what are the obstacles (such as lack of finances or training) and how can these be removed?
- Is there sufficient tourist demand for multiple investment?
- Are there local goods (crafts, food) that can be sold?
- Which uses of the resource are destructive, which are sustainable, and who in the community controls their use?

*Identifying community institutions and leaders*

Chapter 5 emphasises the importance of community institutions, their power, interactions and leadership roles in community dynamics. It is generally better to work with existing institutions and leaders than to attempt to create new groups. However, new issues and problems may require new solutions and hence new institutions.

*Be site specific*

The whole thesis of this book is that the local level matters. Every situation is unique, and although social science methods work everywhere, these methods take time to put into practice and will inevitably come up with different solutions or recommendations at different places (given the different culture, characteristics and context of each place). What is certain is that if the time is not taken at each place to involve local individuals, groups and institutions in information gathering and decision making, then any development project will inevitably fall far short of its objectives.

Section 4

# MODELS OF MANAGEMENT

So far, the book has reviewed social science understanding and methods and their usefulness in examining certain important issues in linking development and conservation through protected areas.

This final section builds on this information in a more specifically management context by examining two models of resource management. The first looks at local level management of resources, and draws on a number of case studies to highlight some of the variety of approaches which have been undertaken.

The second model is that of the biosphere reserve. Biosphere reserves offer significant potential for institutionalising an integrated, local level approach to protected area management. The chapter examines the concept and uses a number of case studies to ground its principles in the reality of existing biosphere reserve management approaches.

# 9 Local Level Management of Resources

One of the most common problems encountered by local people when protected areas are formed has been the restriction of access to resources. Western concepts of protected areas have for a long time been based on exclusion of people. In many cases, local people or indigenous people, who have used lands for very long periods of time, have been arbitrarily barred from certain areas, with little recognition of the ethics, legitimacy or consequences of such actions, on the presumption that protection of lands from local or indigenous people is necessary for conservation. In some situations, there are negative ecological and hence conservation consequences of removing people from lands, although these are often only recognised after some time (West & Brechin 1991).

This is not to suggest that local activities are always beneficial or even benign to conservation. Other change factors – including increasing populations, perhaps due to better health or food services, changes in activities, or declines in resources such as food species because of shrinking habitat – can mean that the effects of local people are no longer compatible with conservation goals. It is in these cases that local solutions need to be developed that take into account the needs and desires of local people and conservation goals. Processes and frameworks that employ social science knowledge, tools and insights can help to achieve this.

In this chapter our focus is on local management of resources. How should an effective participatory model be developed between local people and public and private agencies for sustainable use and conservation of resources? What institutional structures at local, regional and state level best support sustainable development? What policy settings are the most appropriate to support local development and ensure meaningful involvement in, and hence support for, conservation? To illuminate these questions we will analyse a number of case studies. But first we will discuss two issues, common property and institutional linkages, which underpin our analysis of the case studies.

## COMMON PROPERTY

> One of the greatest tragedies of recent times has been the degree to which Hardin's 'tragedy of the commons' has been accepted. (Young 1992: 97)

Many common property systems have worked for millennia without degradation. Aboriginal Australians have managed land sustainably for some 50 000 years. Mongolian herders have grazed the steppes communally for centuries, resulting in an enormously diverse grassland. These common property resources have often been misidentified as 'open access' resources, which are notoriously prone to abuse, but resources managed by a community (communal resources) through custom, tradition, religion and law have often proved very durable, and are especially suited to managing resources that are spatially complex, such as rangelands needed for nomadic grazing and the management of ungulate wildlife.

Sustainable use as a biological conservation strategy assumes that the value of wild resources will foster development that is compatible with conserving biodiversity. That is, if local people are allowed to value wild species for tourism, subsistence, trade and so on, development can occur without sacrificing as much diversity as would be lost if the potential value of the wild resources were not realised.

Central to community conservation is property rights. The three major property right paradigms – community ownership (common property), public or government ownership and private ownership – all have their strengths and weaknesses as a backbone upon which to build successful conservation strategies. These are briefly summarised in the following three paragraphs.

In contemporary societies nature conservation is generally assumed to be a public responsibility as nature conservation is a public or partial-public good (see Chapter 3). Protected areas are almost universally in public ownership and managed by government agencies. There is no doubt that managing a public good like conservation by government on behalf of the public has a simple logic. However, for effective management to occur certain fundamentals need to be in place. First, the government needs to be seen as legitimate by the vast majority of the individuals and communities who make up the nation. Governments must administer natural areas in the best interests of the people. This implies that public officials and government agencies must be committed to the best interest of local communities and biodiversity conservation, and public corruption must be at a minimum. Second, the government must have sufficient administration capacity, infrastructure and knowledge to manage protected areas and wildlife species and to enforce environmental constraints and regulations. Third, the government must be able to raise sufficient revenue from taxes and charges to fund staff, infrastructure, development and ongoing costs of management. If any of these three fundamental requirements are not in place then public ownership of natural resources is unlikely to be a successful means of achieving conservation outcomes. Even in the most rich and developed nations obtaining adequate public funds is difficult. While a high priority may be given to conservation by an individual country's government and its people, an increasingly global economy (with real-time financial markets) is putting pressure on all governments to become more efficient and to reduce government spending. One response to these pressures has been for governments to 'corporatise' resource management agencies to increase efficiency and provide management incentives from use of the resources, e.g. forests, fish, ecotourism and so on.

Private individual ownership of natural resources is widespread in many countries. Private ownership will almost always deliver the most 'efficient' use of the resource (although not necessarily sustainable). It can achieve conservation outcomes in that individual property owners will wish to prevent degradation of the resource to allow for future use and maintain the capital value of their asset. This, of course, assumes a stable economy with low inflation rates. The problems of private ownership are well known. Many of the benefits of a resource, such as conserving the gene pool of a forest for future medical advances, are public benefits which cannot easily be captured by the individual owner, and hence he or she will have less incentive to conserve the forest. Many of the consequences of the production process, such as rising saline water tables on a neighbour's farm caused by clearing a forest area, are external to the production process on the individual's own land and hence there is no incentive to reduce them. Responses to these problems have included: separation of resource property rights into individual bundles such as forestry, agricultural, tourism, mineral, wildlife, all on the same parcel of land; and internalising production externalities by such means as pollution taxes, tradeable permits and so on.

Communal ownership of land, widespread in indigenous societies, has fallen into disrepute in most industrialised developed societies, but its undoubted benefit for nature conservation in certain situations is again starting to be recognised. Common property resources were managed traditionally through custom, religion and traditional law. Where communities are cohesive and strong, and custom, religion and traditional law are in place and appropriate to managing resources communally, then public policy should support such a process as it is a powerful basis for linking conservation and development. Perhaps it can also be a model for people who have lost these traditions to reinvent notions of community at the same time as achieving ecologically sustainable development.

## INSTITUTIONAL LINKAGES

Much has been written about the importance of 'bottom-up' rural development and this has transferred to a new paradigm of community-based or local level conservation. Many community-based organisations (CBOs), non-government organisations (NGOs) and indigenous groups have concentrated on this approach to achieve conservation and development objectives. Agenda 21 clearly recognises that the energy, knowledge, skills and most importantly needs of local people and communities are the fundamental resources that need to be mobilised to bridge the gap between conservation and improving community welfare in an achievable pattern of sustainable development. Ranged against the traditional approaches to rural development and nature conservation, 'top-down' international development aid programmes or central government command and control environment policies, these community-based initiatives have shown a great deal of promise.

However, no policy aimed to support local conservation/development initiatives can work if it ignores the steamrolling impact of the global economy. International trade, the growth imperative, global marketing and communications, instantaneous pricing of commodities, bonds, currencies as well as the power of central governments all provide a context for local processes.

Agenda 21 also clearly recognises this and emphasises the need to mobilise resources at all levels from the local to the global – a partnership between local, regional, national and international stakeholders. It is essential that government policy supports local participation, and at the very least does not inhibit it. There are many failed projects where CBOs and NGOs, funded by international aid, have initiated local participatory conservation/development projects which have not had the support of government organisations and have not made satisfactory linkages with wider political actors or markets for goods, commodities or services. Participation and local involvement are not sufficient.

Scherl *et al.* (1994: 1) talk of a pluralistic planning approach where 'there is an urgent need to link local concerns, local needs and local actions to the national and international government structures designed to conserve and manage the global environment'. We would add there is an equal need in the global development agenda. It is, of course, easy to say this, but how can it be achieved? First, it has to be clearly recognised that there needs to be both genuine community participation and supporting policies and strategies from government agencies and politicians. Initiatives must have clear local benefits and must be financially viable in the marketplace as well as achieving conservation gain. Genuine attempts need to be made by all parties to achieve these linkages. As to how these linkages can be enhanced, several points can be made. The workshop on community conservation at Airlie (The view from Airlie 1993) put forward the idea of 'resource brokers' – external agents, expert in community development and conservation management. We would assert that social scientists with expertise in community structure, social impact assessment, participatory rural appraisal, ethnography, social surveys and policy analysis, both political and economic, can facilitate these linkages.

What we wish to do now is describe some case studies as a way of illustrating the benefits of local management of resources and especially issues related to property rights, institutional linkages and the role of social understanding.

## CASE STUDIES ILLUSTRATING OPTIONS

The literature abounds with examples and descriptive case studies of community management of resources and local development approaches to using natural resources in an ecologically sustainable way. The lengthy history of community forestry, or the extractive reserves model in Brazil and Indonesia, are but two examples.

In this chapter we describe three relatively well known and influential local development approaches to resource management. These cases are then compared to extract common principles that seem necessary for successful local management of ecosystems. The three major case studies are: CAMPFIRE, a community wildlife management programme in Zimbabwe; Joint Forest Management, a joint local community/forest department management programme in various states in India; and Landcare, a community development programme to manage land degradation in Australia.

### Communal Areas Management Programme for Indigenous Resources (CAMPFIRE), Zimbabwe*

The southern African nation of Zimbabwe is famous for its wildlife populations, especially elephant, rhinoceros, buffalo, eland, zebra, waterbuck, kudu and impala. Prior to colonisation in 1890 all land in Zimbabwe was communal, but now state and private land tenure also exists. At present, communal land is legally owned and managed by the state. The state adopted responsibility for the management (including exploitation and conservation) of wildlife resources during the colonial era because of the preference Europeans had for private and state property systems. Indigenous communi-

---

* The CAMPFIRE programme has been reported widely in the literature and this case study draws from many of these reports, particularly Metcalf (1993) and Murindagomo (1990).

ties were excluded from using wildlife resources, as well as, over time, being dispossessed of much of the country's land area. As with many other African nations, the conservation management emphasis followed the model used by the European colonising nations – protected areas which excluded people, and protected wildlife from use or exploitation. Wildlife declined in many agriculturally productive areas, replaced by exotic plants and animals, and habitat remains threatened in other areas because of wildlife's limited usefulness to farmers in the dominant agricultural system. For these reasons, mainly the illegality of hunting, wildlife changed from a useful resource to a problem, or at best an exotic recreational good for an elite.

In addition to the changes in how wildlife was managed, a number of changes in decision-making structures and institutions occurred over many years leading up to independence in 1980. District councils were established in 1984 as a local government administrative unit, with additional decision-making bodies at the subdistrict level (ward development committees and village development committees). These changes were driven by government policy that encouraged decentralised decision making, planning and development. Under the new structure, villages provide direction to wards, who in turn are the political unit of representation from the community to the district council.

A change in the philosophical basis of wildlife management came with a new Parks and Wildlife Act in 1975, allowing landholders to manage wildlife for their own benefit. The changes to this act were prompted by the realisation that the protected area network was in danger of becoming unviable as wildlife declined on surrounding private and communal lands, and that a way was needed to encourage wildlife conservation on these lands. This act paved the way for a different form of management to evolve. The act encouraged private landholders to farm or manage wildlife on game ranches, but it also made provision for district councils to be designated as 'appropriate authorities' for managing wildlife on communal lands.

*The CAMPFIRE programme*

CAMPFIRE is designed 'to give full control of wildlife management to local communities' (Murindagomo 1990: 124). Metcalf (1993: 4) describes it as 'an attempt to make a social link with the economic and ecological objectives of the 1975 Act'. In theory, the programme involves all types of natural resources, although the focus is at present communal wildlife management. The CAMPFIRE model, which was conceived and developed by the Department of National Parks and Wildlife Management (NPWM), had input from ecologists, economists and sociologists, and envisaged rural development practitioners for implementation. The essence of the programme involves the granting of 'appropriate authority' status to district councils by the NPWM. This officially devolves the authority for management of wildlife on communal land to the district council. To be granted appropriate authority under the 1975 act, district councils must have the intention and the capacity to manage resources, with the full participation of the people the council represents, and in such a manner that those people receive any benefits from management. For CAMPFIRE, proof that a council has the support of its membership, a wildlife management plan and the capacity to implement the plan are considered by the department to be sufficient. Because there are few communities with the ability to develop a management plan or the institutional structure capable of implementing one, a body within the NPWM was planned to facilitate and provide technical assistance for this purpose – however, this body has not yet been formed due to problems at high levels. A combination of the Zimbabwe Trust (ZimTrust, a local NGO), the Centre for Applied Social Sciences at the University of Zimbabwe (CASS), and support from NPWM and external agencies such as the World Wide Fund for Nature – Multi-species Project (WWF) has supplied the necessary research and technical support. CAMPFIRE began operating with two councils in 1988–9. By 1990, around 11 districts were becoming involved, and in 1993, Metcalf cited an involvement of 55 districts – almost a quarter of the total in Zimbabwe.

**Plate 9.1** Children warming themselves at dawn in a village in Zimbabwe. CAMPFIRE, a community wildlife management programme in Zimbabwe, has resulted in economic benefits for local people as well as assisting in conservation of wildlife. (Photo: R. Thwaites)

The CAMPFIRE programme is designed to provide a management framework or structure that operates at the local level. As such, specific objectives in terms of conservation are not the object of the programme – these are addressed on a case-by-case basis as the communities involved prepare wildlife management plans. The conservation objectives set at the district level must include a management plan for wildlife (as this is necessary for the 'appropriate authority' status to be granted). Expertise from the groups mentioned above has made this possible. The management plans include quotas for hunting of various species, applied at the ward level.

Two examples of CAMPFIRE, implemented in different areas and different ways, will help illustrate how the programme is operating.

*Guruve District CAMPFIRE project*

Guruve District is located in the northern part of Zimbabwe in the Zambezi Valley. Murindagomo (1990) provides a detailed case study of the Guruve District's involvement in CAMPFIRE, from which the following summary has been drawn. The Guruve CAMPFIRE project was designed for target communities by the NPWM, and did not involve local communities in the project planning stage. The basic aim of the project was to utilise the wildlife resources for safari hunting, with hunters paying to shoot game. This was chosen as the best utilisation of the resources, as the area is remote from tourist routes and therefore unsuitable for tourist viewing of wildlife, which is one of the main forms of capturing economic benefits from wildlife resources. The design of the project involved purchase of equipment for project development (fencing, watering points, offices, vehicles and weapons) and other elements to a total of US$635 400. In addition to the locally managed safari operations, one area is leased to a commercial safari operator. This has allowed the district to compare the effectiveness of local management or a leasing arrangement. The wildlife management project composed part of a larger rural development project (the Mid-Zambezi Valley Rural Development Project). The objectives of this development project relating to wildlife included:

- to conserve the fragile ecosystem and sustain the economic viability of the area through wildlife utilisation;
- to eliminate conflict between agricultural development and wildlife management (through improved crop and household protection);
- to provide increased household income to local people and involve them in the sustained economic use of wildlife;
- to serve as a pilot demonstration for an alternative resettlement model for the drier areas of Zimbabwe;

- to improve nutrition in the area by making game readily and lawfully available to the local people;
- to improve the economic aspects of wildlife utilisation in the area, encouraging more rural communities to adopt it on a commercial scale;
- to improve and master management techniques for communal wildlife management;
- to create local institutions, involving active local participation and communal decision making, for management and development of communally owned natural resources.

*Management and institutions* A natural resources management team from within the NPWM was allocated to the project to provide skills in wildlife management, improvement and marketing systems. The project was designed to use the existing local institutional framework, based on village development committees and ward development committees, which have decision-making, planning and development responsibilities. A district wildlife committee was established, comprised of ward development committee chairs, ward councillors (from wards that had decided to establish communal resource areas) and executive members from the district council. The responsibilities of the district wildlife committee mainly involve the running of safari operations for member wards (employing professional hunters and a project manager) and the management of the benefits of safari operations, which are collected by the committee and distributed to the participating wards. Seven wards were involved in the Guruve district CAMPFIRE project in 1992.

Within the participating wards, wildlife management subcommittees were also formed, to manage, market and conserve the wildlife resources of each ward. Initially, there was no coordination between ward committees, which caused problems and resulted in the need to establish a board of management, consisting of ward representatives, technical advisers from the NPWM, ZimTrust, WWF and CASS. This board coordinates the activities of all the participants and acts as an advisory planning and management body for the district wildlife committee. Support was also necessary in accounting and administrative functions, and this was supplied by ZimTrust.

*Distribution of benefits* When hunters shoot game, returns are required to be filled out and returned to the district wildlife committee. These returns specify the type of game and the ward in which it was taken, among other details. Each ward committee has access to these returns for checking, and each ward committee selects a member to accompany safari hunts in its area.

The basic arrangement is that the wards receive payments for animals taken in their communal resource area. Deductions for operating expenses, as well as a district council levy, are taken from the gross revenue. Meat from sport hunting is also distributed at villages nearest to where an animal is taken, with the remainder available for distribution to wards. Each ward committee also has control over other uses of the communal land, allowing cropping or individual hunting by permit in some cases.

After the first year of operation, the district council, through the district wildlife committee, distributed substantial money to three of the seven wards involved (the other wards had only small wildlife resources). One ward, Kanyurira, with only a small number of households, was able to give direct cash payments to each household – in others, money was used for community projects, given the larger number of households. The difference in benefits experienced by different wards are affected by factors in addition to the abundance of wildlife. Some wards in the district have experienced rapid immigration in recent times after tsetse fly eradication programmes. Other wards, especially Kanyurira, were active in developing land use plans for communal areas, with the help of CASS and the WWF, and were able to make a very strong representation to the council and thus benefit more substantially.

Metcalf (1993) provides figures for Guruve district's revenue in 1992. The district grossed a revenue of US$614 109. The district council levy amounted to US$165 902, while the amount available to wards in total was US$137 696. In the ward that receives most benefit under the programme, Kanyurira, this amounted to US$384 per household. In comparison, the statutory minimum monthly wage is US$40, although many survive on far less in subsistence circumstances (Metcalf 1993). In more basic terms, a goat was worth about US$10 in 1992.

*Beitbridge District CAMPFIRE project*

The Beitbridge District application of the CAMPFIRE concept deserves special mention for a number of its characteristics that set it apart from other programmes. Metcalf (1993) provides a useful case study of elements of this project, which we have drawn on here. The Beitbridge district has implemented CAMPFIRE in a similar fashion to the Guruve district, but has gone beyond the ward level to the village level in differential benefit distribution and in the level of planning and decision making. The most significant difference with the Beitbridge project is that agency assistance has only been in response to requests – a demand-driven process. In order for this to be possible, the responsibility for decision making has been devolved all the way to the village level (by the district council). 'The smallest accountable unit within the district has been empowered and feels secure enough to begin modifying long-standing rangeland practices incorporating a multi-species approach' (Metcalf 1993: 14). Decisions about the balance between private livestock, wildlife resources and communal forage can be made by small groups of households in their planning for the communal rangeland they use (Metcalf 1993). The lack of any overall planned approach (at least by the NPWM) shows that CAMPFIRE can be successful with a much more bottom-up approach than was used in other areas. An initial sum of money was distributed to villages to begin the project – households in one village decided to spend half of their money on a community grinding mill, and keep half.

*Other CAMPFIRE projects*

CAMPFIRE is now being implemented in a large number of districts, in differing ways and with differing success. One district chose not to involve the ward or village level producer communities in the planning, management or operation of the programme (including the management of revenue), depending instead on administrative staff. This application of CAMPFIRE contrasts sharply with the Beitbridge example, and Metcalf (1993: 31) noted that Beitbridge better exemplifies the concept of CAMPFIRE as a community-driven programme.

## Joint Forest Management (JFM), India*

In India, large-scale degradation and destruction of the country's forest resources has occurred over many years, through poor management and lack of control over exploitation and because of the pressures of a growing population. Local communities depend on forests for fuel, fodder, food, fibre, medicine, wood and income and are forced to exploit them regardless of the fact that they no longer have property rights to these resources. The forest departments in the different states of India have attempted to manage forests according to scientific forest management objectives, and in most cases have failed to prevent the (almost total in some areas) decline of forests, with the resultant loss of biodiversity value through plant species loss and faunal habitat disappearance. The extensive logging and clearing of forests that has occurred, especially over the past 30 years, has created vast areas of degraded land, as well as social disruption between local communities and between these communities and the forest departments.

---

\* JFM projects in India, especially in Bengal, are reported widely and we have drawn on many of these reports, especially Campbell *et al.* (1994), Poffenberger (1993), Poffenberger *et al.* (1992a,b), Sarin (1993) and Society for Promotion of Wastelands Development (1993).

An emerging system of management that joins local people and the forest departments in partnership – Joint Forest Management (JFM) – 'holds promise for regenerating India's degraded forests and for empowering local communities in the management of the country's largest remaining land-based common property resource' (Campbell et al. 1994: 2). The JFM model being implemented in many parts of India recognises that sustainable forest management requires the meaningful involvement of local communities. The basic principle of JFM is that local communities will protect and manage forests if they are authorised by the government to do so, if they are able to collect the benefits of doing so, and as long as those benefits outweigh the costs of changing present ways of using the forests.

*Institutions*

One of the main reasons for forest decline has been the change in resource management institutions that have been used to manage forests. Forests were nationalised in the post-independence era, and have undergone a management change from an original 'autonomously devised local resource management system' (Sarin 1993: 1) to custodial management by the state in a uniform, centralised and bureaucratic system. The imposition of a non-indigenous institutional structure, the *gram panchayat* structure, which grouped (often socially unrelated) villages according to administrative convenience, also devalued and weakened local institutions. State control over resource management powers, finances and elections of the *gram panchayats* eroded the legitimacy and traditions of local collective decision making and leadership. These changes in institutions have caused the 'progressive alienation of local people from the forests, weakening or near destruction of traditional community resource management systems, and vast degradation and destruction of the country's forests' (Sarin 1993: 1).

A number of states have been involved in participatory management systems for some time. These systems proved to be very successful in some cases, and were noted by the Ministry of Environment and Forests. The ministry was prompted by this success (and presumably by the crisis obvious in the condition of the country's forest resources) to develop a new national forest policy, released in 1988. This policy called for forests to be managed for environmental protection and conservation, and for meeting the needs of people living in or near the forests. This policy represented a change in direction from a clear revenue-generation philosophy. Revenue and industrial requirements have secondary importance in the new policy, with farm forestry and imports advocated as alternatives to the use of natural forests. The policy also clearly stated the importance of involving forest communities in meeting these objectives. The national government followed the policy document with a notification that state forest departments should develop partnerships with local communities for the protection and management of degraded forest land, and for sharing the benefits of management. Where local people are actively involved in protecting the forests, they may be allowed by the forest department to harvest non-timber forest products for subsistence use, and receive a portion of revenue from the sale of timber when it is mature.

*New institutional arrangements and practices*

Campbell *et al.* (1994) provide a thorough assessment of the changes that will be necessary for the success of JFM. The forest departments must obtain new skills, management tools and better information, as well as the social skills to move from a command and control approach to a participatory and facilitative approach that can cater to the variety of local needs. Substantial changes need to be made to forest department 'orientation, training, internal structure, decision-making processes and priorities' (Sarin 1993: 4). The forest departments will also need to help strong, sustainable and autonomous local institutions to develop (Sarin 1993), a prerequisite for success of JFM, and a responsibility placed on them by the policy and directives of the government. The local communities themselves must recover indigenous knowledge and practices and adopt new skills. Sarin (1993) stresses the fact that state forest departments have for many years had total authority and no compulsion to answer to local people, a situation which will be difficult to change. JFM requires '*mutual* acceptance of clearly defined rights, responsibilities, and accountability by both FDs and local institutions' (Sarin 1993: 4).

In terms of new institutions, communities may need to build new democratic institutions or, where effective traditional institutions exist, to minimise the impact of non-traditional institutions (such as the *gram panchayats*). In doing so, the institutions developed or strengthened must be capable of managing the forest resources based on sustainability, equity and accountability, in a communal system where individual interest is superseded by the communal interest. The following box describes the guiding principles developed by JFM to establish effective local institutions.

**Guiding Principles for Democratic and Effective Local Institutions for JFM**

*Viable, social unit of organisation*
Membership consisting of smaller, socio-economically and culturally homogenous groups such as village hamlets living in close enough proxim-

ity to ensure consistent communication and social interaction tends to be a desirable, though not essential, condition for an effective local institution. The majority of members should have a common interest around which to organise.

*Organisational norms and procedures*

The local institution must evolve norms and procedures, whether formal or informal, acceptable to the majority of members for regulating their actions. These must be based on the principle of equitable rights and responsibilities of all members. Desirable norms include: keeping membership open for all resident adult women and men; effective representation of all minority groups; commitment to promoting gender equity.

*Accountability mechanisms*

For sustainability, each local institution must have effective mechanisms to ensure accountability of individual members and the leadership. These should ideally include: clear and accessible records of accounts and collectively taken decisions; annual elections of formal representatives; participatory and open decision-making with major decisions subject to the approval of the general body; general body power to recall corrupt representatives; penalties for violation of consensual rules.

*Conflict resolution mechanisms*

As periodic conflicts between individual members or different interest groups are inevitable in any collectivity, an effective local institution must have a variety of reliable conflict resolution mechanisms. These could include: strong and trusted leadership; access to impartial, respected outside individual(s) or agencies (e.g. FD, panchayat, council of elders); delegation of arbitration powers to the managing committee or other trusted individuals or organisations.

*Autonomous status*

For any local institution to genuinely function as the voice of its members, it must be an autonomous entity. For effective participation in a JFM partnership, its creation and dissolution should not be controlled by the FD. Both parties may have an equal right to terminate the agreement between them but not to disband the partner institution itself. To enter into a formal JFM agreement with the FD, the local institution should also have an independent legal status.

Source: Sarin (1993: 3)

JFM has seen the development of a number of different local institutions so far that are concerned with the protection and management of forest areas. Sarin (1993) classifies them as:

- groups that have emerged out of local initiative – autonomous village institutions;
- groups promoted and mostly regulated by forest departments under formal JFM programmes;
- government or NGO sponsored development groups that have assumed the additional responsibility for forest protection and/or management.

The first type of group has arisen from local communities through the initiative of local leaders, youth groups, concerned outsiders (including local forest department personnel and NGO members). The success of these groups contributed in part to the change in forest policy of the government in 1988, proving that local communities are capable of effective management action. State forest departments have played little role in their development and continue not to have much influence over their operations. Some groups of this type are small and informal, others larger and more formally organised. In some cases, smaller hamlet-based groups coordinate with each other to regulate boundaries and access controls. Some of these groups have been operating for more than 100 years, while many are much younger.

The second type of group has multiplied as a direct result of the changed policy environment and direction of the forest departments. Forest departments in different states have promoted the formation of JFM partnerships since the policy change (although some cooperative arrangements with local institutions have been operating since the 1930s). Formal JFM programmes have solicited the involvement of local people in forest management, usually in the formation of forest protection committees. These groups are mostly created and regulated by the forest departments (which also have the power to disband them). The forest departments, in setting up groups in such a way, 'show little sensitivity to the basic principles on which democratic community organisations need to be founded and the dynamics of the participatory processes through which they must evolve to be sustainable' (Sarin 1993: 6). Only three out of the thirteen state forest departments involved in forming groups of this type have allowed the local institutions independent legal status.

The third type involves groups that have been formed or sponsored by government or an NGO, with a primary aim of development activities. These groups have had the responsibility for forest management and protection assigned to them in addition to their primary roles.

*PRA techniques*

Participatory rural appraisal (PRA) techniques have been used to assist research involving JFM. They are described in detail in Chapter 4. A formalisation of the research partnerships between local groups, forest departments, NGOs and academic institutions involved in JFM has begun (Campbell *et al.* 1994). Working groups are being set up at different levels in some states – state, circle, division and range levels (administrative levels within the forest departments) – which usually consist of all or some of the following: senior forest department representatives; field level forest department representatives; NGO representatives; and academic institution representatives. These working groups facilitate communication within the department, and between the department and other parties. The main aims of the working groups are to formalise and give direction to research partnerships. An informal national JFM research network also exists, looking at ecological, economic, institutional and training issues (Campbell *et al.* 1994). This network brings together people from state level working groups and nationally based researchers, and also deals with broader policy issues and exchange of information, methodologies and resources. A national support group was formed to coordinate and support this network (within the Society for Promotion of Wastelands Development, an NGO).

The JFM National Support Group has developed social science tools that can be used in JFM, and published them as Field Methods Manuals (Poffenberger *et al.* 1992a,b). This section briefly summarises the methods advocated in these manuals. They include participatory appraisal methods that collect social and environmental information to allow for project development. The PRA technique developed for the JFM programme involved four main categories of information collection:

- community profiling
- determining vegetation changes and condition
- identifying social and institutional management issues
- assessing economic forest production systems

Prior to conducting any research work within the community, the research goals are established in consultation with groups considered to be the target users of the recommendations resulting from the research. These groups included the community members and leaders, local NGOs, forestry department field staff and officers, along with the researchers. This approach ensures that topics of significance to all interested parties may be covered in the research. It also introduces the research project to some key members of the community.

The first step in the research programme is to establish a community profile and determine existing forest management practices. It is important to the success of the research programme that the researchers highlight to community participants that they are the students, learning from the community about issues regarding forest management and resource use within the area. This step involves extensive interviewing of a wide range of community members, such as women, men, children, adults, elderly people and people from various socioeconomic groups. The interviews involve discussions regarding forest and community history, values of the forest, the importance of various forest resources and the way individuals harvest and use forest resources. The community profiling is designed to illustrate the social, economic and religious values ascribed to the forest resources by the community and the interactions between the community and the forest.

Information regarding the vegetation characteristics of the forest is determined using modified and shortened versions of more traditional quantitative research methods. Quadrats are generally used and inventories made, but the community is involved in identifying the species and providing information regarding the history of the sites surveyed. The results from this work are combined with the social and economic data collected during the research programme to formulate site-specific, realistic management options for the forest. Identification and assessment of the social and institution management issues related to managing the forest by the community is critical in the development of strategies to continue local involvement in forest management. In order to assess these issues, the JFM team held further interviews and discussions with various members of the community. Topics included traditional and current management structures and arrangements, conflicts in resource utilisation, identifying institutions for forest management which already exist, identifying key individuals within these institutions, and assessing the incentives and disincentives facing the community in becoming joint managers of the forest.

Among the most significant factors in the consideration of local level participation in forest management is the economic dependence the local community has on forest resources. The community's current use of resources from the forest is determined and the economic returns from this calculated. Such information is gathered through interviews, participant observation and by monitoring the prices of the resources sold at market and time spent harvesting the products. The model of management set out in the recommendations from the research must account for the community's current reliance on forest resources.

Following each field trip, the researchers analyse the data collected. Upon completion of the data-collection process, analysis and interpretation of the data, recommendations are developed with the involvement of the target groups which had input into setting the research goals at the start of the process (i.e. forestry department field staff and officers, local NGOs, representatives of the local community and the researchers). Thus the PRA process has involved the major stakeholders throughout research design, data collection and analysis, drawing conclusions and formulating recommendations. The participatory nature of the process empowered the local communities to highlight issues of priority to them and to discuss these with representatives of the responsible government department. The involvement of the forestry department and the local community in formulating the recommendations from the research allows the implementation of the JFM programme to be site specific, and carried out in the optimal way at each site.

*Case study of JFM in action*

Poffenberger (1993) describes a case study of JFM involving two villages in West Bengal State – Chandana and Harinakuri. The local group involved is of the second type discussed above, but formed initially through the initiative of the local community. The forest area near the villages of Harinakuri and Chandana is about 160 hectares. Two other villages, Nidata and Babunmara, border the northern side of the forest. The village of Chandana, around 38 households, comprises half Bhumi tribal people while the other village people are from scheduled castes. The villagers in Harinakuri are mostly scheduled caste people (around 31 houses). Prior to recent land reform programmes, most villagers were agricultural labourers and tenant farmers, with a significant dependence on the forests for fuel, fodder, medicines, fibres and foods.

The forests were rotation logged during the colonial era by the local raja. Villagers were given access to the trimmings from logging, but were banned from cutting poles or logs in the forest, and the forest was guarded to prevent local use. During the 1950s control of the forests passed from the raja to the West Bengal Forest Department, but the forest was felled completely (including seed trees) prior to the forest department's takeover. Under the administration of the forest department, contractors logged the forest and were accused of failing to leave adequate seed and fruit trees. The contractors were able to do so with impunity, threatening violence to the villagers and department guards if opposed. During the 1970s, political organisers encouraged the local villagers to treat the forests as open areas. In the absence of any managing local institutions, villagers and contractors began uncontrolled exploitation of the timber resources, leading to very severe degradation by the 1980s. Repercussions, including shortages of wood for making tools, changes in hydrology leading to wells drying up more quickly, and (based on anecdotal evidence) diminished rainfall and hotter temperatures, became apparent.

With this background of changing and ineffective institutions and social change, and the consequent decline in the values of the forests, a local villager from Harinakuri began visiting Chandana to discuss forest management problems. This man, with only two years of formal education and formerly a landless labourer, had been granted land through the West Bengal State Government as part of a land reform programme. He felt that it was important that the community should control forests as they were now able to control agricultural land. After visiting village houses separately, speaking about the consequences of forest degradation, village meetings were organised. By 1984, Chandana village and three neighbouring villages met and decided on a collaborative protection programme, where each community should take responsibility for the area of forest nearest their village. Boundaries were defined, and Chandana and Harinakuri villages set up an active protection programme. The other two villages are less active in protection, although their involvement is important as they border the other side of the forest. The four villages form a forest protection committee, under the JFM programme of the West Bengal Forest Department.

Protection of the forest involves keeping people from other villages (who come to cut fuelwood for sale for cash income) out of the forest. In many cases, higher income families in other towns and villages pressure tribal and lower caste members to cut wood for them on a contract basis. At some times of year, larger groups come to cut poles for commercial sale. During these times, a patrol is kept up. If the Forest Protection Committee catches men cutting poles, they levy a fine and confiscate axes, or turn them over to the forest department guard, and they are later fined. In some cases, patrols of armed villagers are necessary to keep out cutting groups.

The costs to the villagers of foregoing income through fuelwood cutting and timber harvesting is substantial for these villages. The time spent in patrolling is also considerable. Villagers knew, however, that exploitative activities were eventually going to lead to a loss of the resource and they would eventually need to find other occupations. This knowledge has helped villagers to make the decision to change their behaviour. The land reform programme that transferred land title to previously landless workers also made it possible for this change to be made, as villagers no longer have to give substantial proportions of the food they produce to a landlord. Some villagers have found other sources of income, such as production of puffed rice for the local market.

The forests in the area that are under protection have recovered rapidly. Many flora and fauna species have returned, and timber stands have increased rapidly. Gains in the form of non-timber forest products, such as tubers, mushrooms and fibres, have also been made. Slowed runoff, improved groundwater infiltration and an increase in bird species which control crop pests are other benefits of protection. Harvesting of the timber is pending, and the villagers to some extent have different thoughts to the forest department about when and how the timber resources resulting from the protection should be exploited. Some villagers were keen to harvest very early, feeling the threat of a mass effort by people from other villages to take the timber. Although support existed among the villagers for a rotation cutting regime, many supported a total felling for the first time at least, to prove to outside villagers that they should be protecting their own forests by demonstrating the financial benefits of protection in a dramatic way. The forest department supported a 10% per year felling regime, to provide a more sustained income and to ensure the non-timber benefits are retained. The forest department policies allow for the forest protection committees to receive 25% of the net income from collection of timber. This contrasts with the impression the villagers had that they would receive 40% of the gross. Overhead costs taken by the forest department in other areas are high, and so income could be substantially less than villagers expect. The forest department is addressing the issue of overheads, but communication with the forest protection committee about information on possible income has been poor to date. The villagers have decided to maintain the revenues they will eventually receive as a community fund and want the forest department to establish a bank account for them, overseen by the forest protection managing committee. Funds will be used to construct a community rice storage barn, which will help the families involved in the puffed rice production. A cooperative store to sell groceries, stationery and school supplies, and a savings and loan programme for community members, are also desired. For these to occur, the forest protection committee will need to become a formal legal entity, which may require the help of the forest department.

*The success of JFM*

Campbell *et al.* (1994) noted that 15 of India's states have passed resolutions and orders initiating JFM activities. Over 10 000 formal and informal local groups are involved in forest protection or management, covering around 1.5 million hectares of forest (2% of the forest area of India). 'Communities are taking up the challenge of protection, accepting the opportunity costs of regulating use patterns, developing democratic processes to ensure equity in decision-making and the sharing of benefits, and struggling with inevitable conflicts over access to regenerating resources' (Campbell *et al.* 1994: 3). The success of an operation that has involved such large numbers of communities is obvious. The ability of JFM to be able to cope with the variety of local conditions and situations remains to be seen, and some concern exists that the 'conventional, top-down forest planning and management may be extended to an equally centralised and inflexible prescription of rules and regulations which direct the creation, constitution and functioning of community organisations expected to participate in JFM' (Sarin 1993: 6). Many authors stress that the diversity of different local situations and institutions means that JFM must be responsive at the local level, 'tailored to respond to prevailing problems and opportunities' (Poffenberger et al. 1992a: 3). The wide variety of different areas and communities involved so far seems to indicate that this should be possible.

### Landcare – a Community-based Programme for Sustainable Land Use in Australia*

Landcare is a participatory land conservation programme operating on a large scale throughout rural (and, on a lesser scale, urban) Australia. Landcare involves people forming local level groups to address land conservation issues of all types. Australia has suffered vast land degradation in the little more than 200 years since European invasion. Soil erosion, salinity, acidity and other soil problems, water quality reduction and river and wetland contamination, clearing of tall forests and woodlands, extinctions of many plants and animals, and huge pest animal and plant problems have been the result of European presence and practices. The enormous cost has been not only in terms of loss of conservation values, but also in loss of economic and agricultural production, with the consequent social costs. Accompanying the severe problems affecting agricultural land (caused by a combination of different factors, but predominantly by the use of inappropriate techniques imported from Europe) has been a severe rural decline, with many farms operating at a consistent loss in consecutive years, resulting in large average losses in many farming sectors. Poverty and hardship in Australian rural areas have their obvious environmental consequences – 'more sustainable systems of land use and management are unlikely to be developed or implemented by people preoccupied with short-term survival' (Campbell 1994: 21). It is with (and perhaps despite) this background that Landcare has developed in the different states of Australia.

*Background and development*

The landcare programme emerged as a distinctive entity in the state of Victoria during 1986 and has been embraced by governments, farmer organisations and conservation groups throughout Australia as offering a model for effective community action to manage land degradation and assist the

---

* Landcare has been widely reported in the literature. As well as this literature this case study draws on the research done by the authors in collaboration with colleagues in the Johnstone Centre coordinated by Allan Curtis. See also Chapter 4.

move to more sustainable resource use. Indeed, by January 1995 there were an estimated 2200 Australian landcare groups involving 30% of the farming community (Mues *et al.* 1994), including 500 Victorian landcare groups, with an average of two new Victorian groups formed each week for the previous three years (Curtis & De Lacy 1995). After lobbying from major farmer and conservation groups, the commonwealth government committed spending of $360 million in the 'Decade of Landcare' programme announced as part of a major environmental statement by the Prime Minister (Hawke 1989); landcare was now a national programme. Landcare is intended to achieve more sustainable use of Australia's farming lands (DCE 1992) and to enhance biodiversity (Farley and Toyne 1989). Whilst governments espoused a 'landcare programme' that embraced all facets of sustainable resource use, it was the emergence, growth and activity of these voluntary groups which captured public attention and distinguished landcare from previous efforts to achieve more sustainable resource use.

**Plate 9.2** Ranger helping local school children to plant tree seedlings. Landcare, a participatory land conservation programme, facilitates involvement of different groups in local environmental projects. (Photo: R. Thwaites)

Most landcare groups have developed in rural areas and group membership is voluntary and open to any member of the local community. Groups frequently operate at catchment or subcatchment scales and are involved in a variety of activities related to the management of issues such as water quality decline, soil erosion, ground water salinity, soil acidity and vegeta-

# LOCAL LEVEL MANAGEMENT OF RESOURCES 201

tion decline and introduced pest animals and weeds. Amongst their various activities, landcare groups hold meetings to discuss issues, identify priorities, develop action strategies and debate a range of resource management issues; conduct field days and farm walks and establish demonstration sites; undertake a variety of educational and promotional activities such as hosting tours and involving other community groups in landcare activities, organise conferences, write newsletters and field guides and prepare media releases; undertake a range of on-ground work including seed collection and tree planting, constructing salinity and erosion control structures, coordinating pest animals and weed control, erecting fencing to control stock access to creeks and streams and establish wildlife corridors; coordinate planning activities related to property and catchment planning; and some members are involved in the preparation of submissions for government funding.

Campbell (1989) and Curtis *et al.* (1993) discussed the benefits of landcare participation from the landholder perspective in terms of: landholders being able to share problems and ideas and in doing so gain support and encouragement to push ahead; working together to tackle common problems more effectively; having opportunities for learning about land management and to plan at property and catchment levels so that resource management is based upon a shared understanding of important physical, social and economic processes operating within and beyond the farm gate; obtaining financial and technical assistance from government which they would be unlikely to receive as individual landholders; and having increased opportunities for social interaction with other members of their local community.

To a large extent, enthusiasm for landcare can be attributed to widespread acceptance of the view that traditional approaches to agricultural extension involving the transfer of technology from government officers to individual land managers have failed satisfactorily to affect landholder perceptions of issues or result in widespread adoption of more sustainable practices. Hartley *et al.* (1992) suggested agency support for landcare representing the adoption of a new participatory model for agricultural extension where agency staff empower or assist farmers to manage change and solve problems through group action. Woodhill *et al.* (1992: 263) suggested landcare involved 'community development' which is the prerequisite for 'participative approaches to land conservation to be effective' and explained that 'by community development we mean building within local communities the understanding, commitment, knowledge, skills and resources to effectively engage in a long term process of developing sustainable land use practices'. The landcare approach offers a means of tapping indigenous farmer knowledge (Chambers 1986) and would appear to fit Roling's (1988) requirements for effective agricultural information systems in that it attempts to work with all landholders rather than expert farmers, to develop strong utiliser constituencies and has a strong focus upon human resource development.

A case study of the formation and activities of one landcare group will show some details of how the programme operates in practice.

*Kalannie-Goodlands Landcare Group*

Campbell (1994) provides a comprehensive case study which has been summarised here. The Kalannie-Goodlands area in Western Australia originally supported a very wide diversity of native plants and animals, and a vegetation mosaic of heath, mallee and woodlands. This landscape has suffered almost total clearing, much of it in the latter half of the twentieth century, and the area has been largely changed into a monocultural landscape of wheat, lupins or annual pasture species grazed by sheep. The area now has an extremely high number of endangered plants and animals. A Land Conservation District Committee was formed in 1988, due to concern about waterlogging, salinity, erosion and surface runoff. Around 100 members are involved in the group and the group area covers 300 000 hectares, much of which is degraded. The group has undertaken a major project to prepare a district plan, to ensure that the actions of individuals contribute to the overall welfare of the environment at a catchment level. Identification and quantification of the major land management problems have formed the first part of the process. The planning is supported by national funding, a local levy, and by expertise from employed consultants and agency staff. Property management plans are also being developed by individual farmers with the ultimate aim of developing sustainable agricultural systems for properties and subcatchment areas. Workshops to help in developing these plans were arranged by the group. The group investigates salinity control options, and sets up demonstrations of tree plantings. Seeds from tree planting are collected locally and propagated by group members.

Some economic benefit for local people has come directly from the landcare programme. The need for property plans has been highlighted by the group, and one local person has established a Geographic Information System business to service the group's members. The propagation of a large number of trees for landcare group plantings and for members to plant on their properties has also become a profit-making exercise for some members in the landcare group.

The group has structured itself with a committee and an executive committee, and has developed a cabinet portfolio system where individuals become responsible for group activities of one type – such as tree planting or newsletter production. One concern of the members is that many of the planning solutions available are capital intensive, therefore the implementation of the property plans that the group members have developed will cost a great deal of money, which is unavailable to a rural community in crisis. A vision is expressed by one group member, however, of a landscape eventually changed through group action, from broad areas of bare land to multiple rows of trees meandering through the landscape, lowering water tables and preventing them from rising, providing for wildlife, with control of salinity and erosion.

*Social research – landcare evaluation*
Curtis and DeLacy (1995, 1996a,b), using social research methods, undertook an extensive nationwide evaluation of landcare which is briefly reported here.

Evaluators can turn to a number of sources in their effort to unravel programme theory: they can approach programme staff, clients and other stakeholders for their views; they can review literature on the programme under scrutiny or similar programmes; they can examine programme documentation; and they can observe programme operation. Given the lack of explicit programme goals, the diversity of stakeholder opinions about landcare programme objectives, and the heterogeneity of programme implementation at the local level, the authors devoted considerable energy to unravelling programme logic. To a large extent, this task was accomplished using information collected through the state-wide group activities report process, secondary data such as government policies and strategies, the authors' knowledge of landcare group activities in northeast Victoria and personal contacts with stakeholders at local, regional, state and federal scales. The authors determined that the key assumptions underlying the community landcare programme (Figure 9.1) were that with limited government funding of a self-help programme, landcare group action will facilitate a process of community participation that will mobilise a large proportion of the rural population and produce more aware, informed, skilled and adaptive resource managers with a stronger stewardship or land ethic, and thereby result in the adoption of improved management practices and assist the move to more sustainable resource use.

*Methodologies* A variety of quantitative and qualitative research instruments were used to evaluate the effectiveness of the community landcare programme using the programme rationale as the focus. In collaboration with lead agencies, a majority of Victorian landcare groups were surveyed during 1991/92 and 1993 to explore programme implementation, assess group effectiveness and gain insight into the agency/community relationship and the role of women in landcare. Similar surveys were undertaken in the states of Western Australia, Tasmania, South Australia and Queensland, providing an Australia-wide sample of 471 groups.

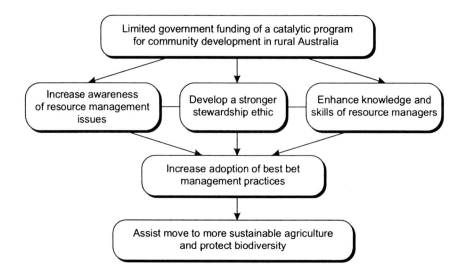

**Figure 9.1** Model of the programme logic for community landcare in Australia (Curtis & De Lacy 1996b)

Until recently, landcare groups had not participated in any real sense in landcare policy development and it was in this context that the authors investigated Victorian experience with community reference groups developing nine regional landcare plans (RLAPs). These plans formed the basis of the state's response to the commonwealth landcare initiative and the framework of legislation for integrated catchment management. Participant observation of the RLAP process in northeast Victoria during 1992/93 and a mailed survey using a mix of closed and open-ended questions to all Victorian RLAP participants during 1993 enabled an assessment of the extent to which key stakeholders were represented and had contributed to landcare strategy and policy development.

A survey of all rural landholders in 12 subcatchments of northeast Victoria was undertaken during 1993 to describe the characteristics of landcare participants and non-participants, explore reasons for participation and non-participation in landcare, and assess the impact of landcare upon key programme outcomes. Whilst most listed rural property owners were men, with the assistance of local residents the researchers were able to target 20% of surveys to rural women. As part of this study, the authors conducted a series of focus groups with small numbers of women who were landcare participants to provide further insight into the impact of gender upon landcare participation and experience.

*Findings* The research revealed the vast scale of community participation in landcare and established that landcare had moved extension beyond the 'expert farmer' group. Landcare groups and landcare participants had undertaken considerable on-ground work likely to assist the move to more sustainable agriculture. Information from the northeast Victorian landholder survey indicated that landcare participation had significant impact upon landholder awareness of issues, level of knowledge and adoption of 'best bet' practices. Groups had successfully 'pulled down' resources for landcare work and landcare activity appeared likely to provide additional political leverage for local communities. This information suggested that landcare group activity had made an important contribution towards sustainable resource management and was likely to endure. However, research findings also suggested a number of flaws in programme logic. Given low levels of profitability amongst landholders, the vast scale and intractable nature of key issues and the considerable off-site benefits of remedial action, it is debatable whether limited funding of a communication process will effect behavioural changes sufficient to make a difference at the landscape level. Programme emphasis upon developing landholders' stewardship ethic also appears misplaced: there was no significant difference in the stewardship ethic of participants and non-participants. Indeed, to the extent that landcare focuses upon changing individual behaviour rather than societal barriers to rural development, it is open to the criticism that it places too much responsibility upon individual landholders. This research confirmed the community/government partnership as a fundamental element of landcare. Agency staff had regular contact with most groups, provided valuable information and assistance, and most groups received government funding that appeared crucial to group work. There was evidence that landcare participants contributed to policy development and institutional arrangements for power sharing were emerging. Whilst it is inappropriate to assess landcare by the standards of spontaneous and autonomous community participation, government appeared to have greater influence upon landcare decision making than is consistent with programme rhetoric.

## COMPARISON AND ANALYSIS OF THE CASE STUDIES

The three case studies vary widely in geography, culture, stage of development and conservation problems. However, analysis of the three approaches highlights some interesting similarities. Table 9.1 compares some of the characteristics of the projects and analyses them in terms of policy context, local institutions and types of linkages.

Table 9.1 Comparison of characteristics of case studies

| | CAMPFIRE | JFM | LANDCARE |
|---|---|---|---|
| Resource use | Harvesting target species | Timber harvesting | Use of land resource for agriculture |
| Resource type | Threatened wildlife resource | Varies – at worst, badly degraded or non-existent forest resource | Varies – at worst, badly degraded land resource base |
| Broad conservation objective | Reduced threat and active management for wildlife population growth | Restoration of forest resource | Restoration of land resource |
| Broad development objective | Income for local villagers, control over communal resources, change from unsustainable use of resources to more sustainable wildlife resource use | Income for local villagers, change from unsustainable resource use to sustainable resource use | Increased capital value of asset, change from unsustainable resource use to sustainable resource use |
| Political-policy context | Supportive policy for local involvement at national level (legislative), recent changes to political structure at local and district levels | Supportive policy for local involvement at national level but not in all states (legislative) | Supportive national and state policies for local participation |
| Institutions | Local institutions disempowered first through colonialism, but local authorities and village committees reempowered on independence. Programme makes use of these existing institutions | Variety of local institutions involved, some in existence for a long time, the majority newly formed specifically for JFM | New institutions formed for landcare programme |
| Local involvement | Local people involved at differing levels (district, ward or village) in some cases locally operated and controlled | Local people in some cases instigated forest protection activities (bottom up), in other cases departmentally instigated (top down). Some local institutions have autonomy, others do not. Local institutions generally do not have complete control over benefits or management | Programme initially introduced in a top-down fashion. Groups form through local initiative, operate autonomously and draw down government programmes as appropriate |
| Need for continuing non-local input | Operations in many districts are able to be district driven and managed, though require support from government agencies | Partnership arrangements mean that control will always be shared with forest department's local groups | Local groups autonomous but require coordination input from government agency. Networks and links with non-local institutions remain important |
| Conservation/development linkages | Sustainable wildlife harvesting is legal and economically viable | Conservation of the forest is essential for continued use of the resource | Conservation work provides long-term environmental benefits along with financial gain through improved productivity and capital value of the land |

The similarities seen in these three cases may, of course, not be generalisable. The cases were chosen mainly because there seemed to be widespread agreement by practitioners and in the literature that they were all important policy initiatives having significant impact in their own country as well as a wider influence internationally in attempts to link development with conservation through local level action. Consequently, we would suggest that the issues raised in this comparison may indeed highlight important principles. We go on to list five such principles.

First, and most obviously, all three highlight the importance of community action and *participation*. Although the strength of participatory action varied from site to site within each case study, we are left with a strong sense of all three being community owned.

Second, all cases emphasise the fundamental importance of *linking* conservation objectives to clearly perceived development outcomes. In each case local individuals and groups saw clear development benefits from the policies which directly achieved environmental improvement. These development benefits were numerous, but in all three cases included specific financial gain.

Third, to achieve development and conservation linkage often requires careful consideration of developing appropriate *property right* regimes for the resources in question. In all these cases full or partial property rights for the resource (wildlife, forests, agricultural land) resided with local individuals or community groups. And furthermore, reform of these property rights, from state to community ownership, was a crucial factor in CAMPFIRE and Joint Forest Management.

Fourth, the need for *building* local community-based *institutions* to manage the resource was important in all three cases. Interestingly, all three required the establishment of new institutions as well as strengthening existing ones. Much of the development literature highlights the preference for using existing institutions over creating new ones. These three cases could perhaps question how important this is.

Fifth, in each case, local community action, development and management was clearly supported by public policy and linked to individual government agencies. Whether the initiatives for implementing the programme were 'top down' or 'bottom up' seemed to vary from site to site within each broad policy implementation. But we are left with an overwhelming sense of *partnership* as a crucial common factor in the success of all three programmes.

## CONCLUSION

Our analysis of some 50 conservation/development case studies from around the world highlights the importance of explicitly linking individual and local community benefit to conservation programmes. If conservationists stick to the old paradigm, that development and conservation are antithetical and natural resources are for preservation not use, then our review of case studies holds out little hope for real conservation gain. Our analyses of the case studies also illustrate the importance to achieving successful conservation outcomes of having suitable property rights regimes. Individual or community ownership of resources is needed to ensure some equity of power relationships between the local and the centre, which is essential for successful negotiation and participation to occur. Policy reform, to achieve a better regime of resource and property rights, is one of the most pressing issues in achieving biodiversity conservation. Finally, it is crucially important that the policy settings from state and central government are supportive of locally managed resource programmes. Successful community conservation programmes are the result of partnerships between government agencies and local institutions – a meeting of 'top down' and 'bottom up' in a genuine participatory, collaborative programme.

It must be emphasised again that not only do people need development, but the irony is that in our modern world driven by market economics, real-time individual and mass communication and global marketing, nature conservation also needs development. Or, more correctly, if local communities can benefit from nature conservation, then biodiversity is likely to be conserved. If nature conservation policies do not allow for local development, the resource will be degraded.

# 10 Biosphere Reserves

In 1968 UNESCO organised what became known as 'The Biosphere Conference', a landmark meeting which eventually recommended an international research programme on 'Man and the Biosphere' (MAB). This conference saw the beginnings of the programme. One of the early outcomes of the MAB programme was the establishment in the early 1970s of an international network of biosphere reserves.

The Biosphere Conference is now recognised as the first public forum where sustainable development concepts, as they are currently accepted, were discussed. The MAB programme, as initiated at the conference, directly promoted the recognition of the development/conservation nexus, and its importance to the sustainable use of the world's resources. The biosphere reserves established by the MAB programme provided models whereby environmentally sound and sustainable development could be promoted in areas adjacent to the more strictly protected areas.

Reviewing the programme 25 years later, Batisse (1993: 108) commented that:

> the single most original feature of the Biosphere Conference was to have firmly declared, for the first time, that the utilisation and the conservation of our land and water resources should go hand-in-hand rather than in opposition, and that interdisciplinary scientific approaches should be promoted to achieve this aim.

By 1981, 208 biosphere reserves had been designated in 58 countries. Although the three core roles of biosphere reserves (the conservation role, the research and monitoring role and the development role) were articulated, the first 10 years of the programme saw the creation of many reserves based primarily on ecological criteria, in many cases simply overlaying existing protected areas. Any extra activity (over and above what was already in place in the pre-existing protected area) was mostly associated with research in the natural and physical sciences. The social/human portion of the MAB concept, the development role, had little expression in the designation and functioning of the early biosphere reserves. Because the people and community development aspect of the concept was diluted, in nearly all cases the envisaged social science research, so necessary to understanding the link between development and conservation, was never carried out.

With the increasing consciousness arising from UNCED and Agenda 21, there has been a renewed interest in the biosphere reserve model as a framework for integrating local development with biodiversity conservation. We will in this chapter look in depth at the model and especially consider some case studies where the conservation development nexus has been highlighted and social research methods used to assist in its implementation.

## WHAT ARE BIOSPHERE RESERVES?

Three main roles for biosphere reserves have emerged, along with a model for the zonation which allows for differing degrees of human intervention in various areas. The zonation proposed before the first biosphere reserves were established remains today in the form of the core, buffer and transition zones of the reserves. At the Minsk congress in 1984 a world 'Action Plan for Biosphere Reserves' was drawn up and later adopted by the MAB Coordinating Council. Under the recommendations of this action plan, the worldwide biosphere reserve network expanded to 324 individual reserves in 82 countries by March 1995, when the Seville conference was held to renew the action plan.

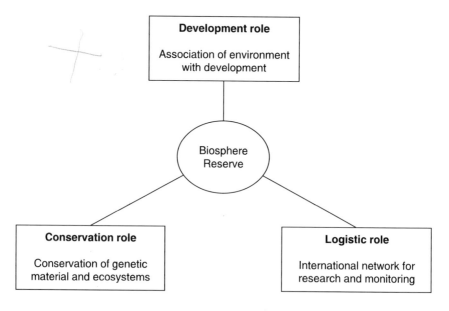

Figure 10.1   The biosphere reserve model (after Batisse 1986)

The three main roles of biosphere reserves, as shown in Figure 10.1, are described by Batisse as follows (1986):

- a conservation role – providing protection of genetic resources, species and ecosystems, on a world-wide basis;
- a logistic role – providing interconnected facilities for research and monitoring in the framework of an internationally coordinated scientific programme; and
- a development role – searching for rational and sustainable use of ecosystem resources and hence for close cooperation with the human populations concerned.

Figure 10.2 shows the biosphere reserve model envisaged as an abstract theoretical construct. This particular diagram shows a complex multicore reserve plan. Although this ideal geographical and functional form of a reserve is rarely achievable, the model shown here gives a diagrammatic and conceptual baseline which can provide an understanding of how functions and zones are placed within the biosphere reserve model.

Basically, the biosphere reserve model includes a core zone (with its functions of conservation and monitoring), a buffer zone (with functions of research, education and tourism) and a transition zone (which functions as a zone where influences and activities of people must be understood and taken into account for their impact on the resource, and therefore where the processes and difficulties of integrating conservation and development may be examined and understood).

The core area of a biosphere reserve is strictly protected, managed via well defined conservation objectives, and represents typical samples of natural or minimally disturbed ecosystems. The core area (or areas) should be large enough to act as *in situ* conservation units and provide an area for baseline monitoring studies of the ecosystems of which they form part. In the ideal model, the core area should not have any significant human settlements, but may be subject to different types of management according to the conservation objectives. The core area may also be undeliniated, existing as the inaccessible part of a delineated buffer zone. The core area should be used for non-destructive research, as well as environmental observation and monitoring.

## Buffer and Transition Zones

The buffer zone of a biosphere reserve surrounds the core area and, as mentioned above, serves to insulate the important core areas. The buffer zone should be well delineated but may contain an undeliniated core, with both being managed as a single administrative unit. Many national parks already in existence fit this model, with wilderness or reference areas surrounded by less strictly protected areas within one park. Ideally, buffer zones need to have clear legal status (although often more than one administrative body will be involved), and only activities compatible with the protection of the core zone should take place. These activities might include research, environmental education and training, ecological farming and tourism. The buffer zone may also protect areas of land for future research needs.

The transition zone, or zone of influence, of a biosphere reserve often supports human settlements and a range of economic activities. Actions in this area impact on the core and buffer zones of the reserve and, as such, are monitored to encourage environmentally sound activities. The zone of influence is where knowledge gained from the core and buffer zones is implemented to achieve sustainable development.

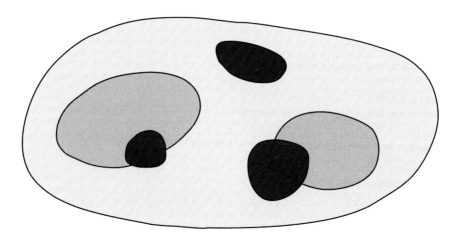

Figure 10.2   Cluster biosphere reserve model (after Batisse 1986)

*Importance of buffer zones*

The biosphere reserve model rests heavily on the concept of buffer zones. Although the term is used very specifically in the biosphere reserve model, buffer zones have been used in differing ways for many years. At this stage we will make some comments about buffer zones in general, as in some cases the use of buffer zones as a management strategy can mimic aspects of the more comprehensive biosphere reserve model. Buffer zones can provide a transition between areas used primarily for conservation purposes, and areas that are used for purposes not well suited to conservation (such as intensively farmed land). We will adopt Sayer's definition (1991: 2) of a buffer zone:

> A zone, peripheral to a national park or equivalent reserve, where restrictions are placed upon resource use or special development measures are undertaken to enhance the conservation value of the area.

Of particular importance (for our purposes) is the part of the definition recognising that development activities may take place. Buffer zones are (and have the potential to be) ideal sites for integrated conservation and development projects, given that local people are often restricted to areas close to parks and that impacts in these areas can be most pressing. Sayer notes that buffer zone projects have also tended to attract funding from international donors wishing both to become involved in nature conservation and to benefit local people. While many projects to establish buffer zones and develop ecologically acceptable buffer zone activities have been undertaken in the last few decades, in general the assessment of experts in conservation is that these projects have been mostly disappointing. Despite the problems of establishing buffer zones as aid or development 'projects' with all the problems that are associated with this perspective, there are many benefits that buffer zones can provide. Sayer discusses these (1991), which include biological benefits and social benefits. Biological benefits include: providing a barrier preventing access to the core area and discouraging illegal resource use; protecting the core area from exotic species invasion; providing some protection against storm damage, drought and erosion; and providing more habitat for species that range widely. Social benefits include: the presence of a flexible mechanism for resolving land use conflicts involving conservation; the provision of some benefits similar to those of which local people may be deprived by the protection of the core area; the improvement of the local environment; increased earning potential of local people; support for conservation that may be generated locally or regionally; protection for the traditional rights of local people; and the provision of a reserve of species for restoration purposes.

Sayer (1991) cites a number of useful illustrative case studies, two of which we will mention briefly to demonstrate the nature of buffer zones and activities.

### Forest Reserves in the East Usambara Mountains in Tanzania

These preserve the remnant vegetation areas for that area, and host many endemic flora and fauna species. These reserves are subject to pressure from illegal agricultural activity, timber harvesting, and cardamom cultivation from the local people. The large rural population, many of whom suffer from the depression of the dominant tea-growing industry, are forced to utilise forest lands and products. A project to improve the living conditions of the local people and preserve the biological diversity and environmental value of the forest has been undertaken by the IUCN and Tanzanian governmental bodies. Rural development activities in the zone surrounding the reserves have been designed to remove the need for local people to exploit the protected areas. Woodlots and tree plantings, some of wood-producing species and others of cloves, pepper or coffee trees, have been encouraged. Pit sawing of timber in the forests by non-locals has been the main method of illegal timber harvesting, and the project has also tried to develop ways of placing this form of resource use under the control of local people. The development activities are managed through village coordinators, who are local people trained by the project to help facilitate the formation of village development committees. These committees are the means by which the project provides support for village initiatives. The project has been successful in facilitating the planting of around half a million trees, including boundary strips and communal plantations. A number of conclusions have been drawn by the agencies involved and the project staff which are interesting in terms of the social science context: the short funding commitment has made long-term planning difficult; local community attitudes and behaviour have been found to take a long time to change; a long period of time was discovered to be necessary to develop a useful and meaningful relationship with local people; planning for rural development was not in this case very useful, with the best ideas instead emerging from discussions or experimental plots; salary supplements were necessary for government employees to make progress on the project; and the village coordinator system was very successful, making a major contribution to awareness of conservation issues and providing a good foundation for further development work.

**Cross River National Park in Nigeria**

This is a newly formed protected area for moist forest conservation located in the southeast of the country. A project to design, establish and develop the new area was set up in 1988, involving WWF, on behalf of the Nigerian government and the Commission of the European Communities. A large number of villages surround the park, with around 76 000 people living nearby, many of them dependent on the park for their income and subsistence through hunting, fishing and collecting forest produce. Most of these people live by shifting agriculture, and the park is threatened by the continuation of these activities. The project defined a support zone (feeling that this implies mutual support rather than a defensive relationship with the local people) containing target villages, farmlands and communal forests. Within this support zone, the project intends to provide incentives to encourage local people to participate actively in the protection of the park and to move to more sustainable land use systems. The support zone is defined by the level of dependence which each village has on the park, rather than on a geographic area. The incentive system will involve the administration of credit, advice, grants, planting materials and other benefits. These are conditional on villages respecting the boundaries of the park. Local people only will be eligible for benefits, and the benefits will be of a fixed sum nature for villages which may discourage registration of non-locals. The unit of management at the local level is the village council, but a village liaison officer will be employed in each village to facilitate communication. The project will also be able to sanction individuals or villages for non-compliance with park rules, including the refusal of loans and withdrawal of resource use privileges.

## The Function of Biosphere Reserves

Parker (1993) describes four main functions of biosphere reserves – conservation as an open system, education and training, research and monitoring, and cooperation.

The *conservation as an open system* function is important to the biosphere reserve concept. This function rests on the premise that conservation should occur in an open setting – the reserve is basically an area of ecologically representative landscape where land use is controlled but can range from strict protection to intensive sustainable production. The conservation function of reserves embodies genetic resource conservation (as a source for potential use or for reintroduction of species to other areas, for example), as well as the conservation of cultural elements such as traditional land use systems which illustrate the relationship between people and the environment. In terms of the objectives of biosphere reserves, these indigenous land use systems may be of immense importance in the search for sustainable ways to use and manage land.

The *research and monitoring* function is significant also in biosphere reserves, especially as core areas are often ideal as sites for monitoring change (due to their size and freedom from substantial impact). The function of biosphere reserves as sites for multidisciplinary research (using natural and social sciences) to develop models for sustainable development over a wider region is also very important but, as mentioned earlier, underutilised. Biosphere reserves can be sites for a wide range of research: determining requirements for conserving diversity; assessing impacts of pollution on ecosystems; evaluating the effects of traditional and modern land use practices on ecosystems; developing sustainable production systems for degraded areas; and investigating methods and requirements for sustainable living (among others). The research and monitoring function also includes the international linkages for comparative studies in different parts of the world, for testing, standardising and sharing methodologies, and for coordinating the development of information management systems.

The *education and training* function of reserves is facilitated by their ability to act as field centres for educating and training scientists, managers, administrators, visitors and local people in conservation and protected area issues. The different reserves may be able to offer different opportunities, although general academic and professional training, environmental education, demonstration and extension, and training for local people are desirable. Tourism may also play an important role in these education functions.

The *cooperation* function of the reserves is fundamentally important (and an important way in which biosphere reserves differ from other protected areas). Cooperation is embodied in the concept of biosphere reserves and has symbolic as well as practical value, encouraging personal commitment in people. Cooperation at local and regional levels is diverse and brings together people with a range of interests, directing efforts towards strategies for dealing with environmental, land use and socioeconomic problems affecting a particular region. The cooperation function may also bring together government agencies and academic institutions to provide a perspective on regional problems of ecosystem use and management.

## THE SEVILLE CONFERENCE

In March 1995 UNESCO held an international conference on biosphere reserves in Seville, Spain. The aim of the conference was to establish the basis for a new action plan for biosphere reserves into the twenty-first century. The conference was attended by reserve managers, reserve network coordinators, natural resource management policy makers and scientific researchers, all of whom contributed viewpoints on how to improve on the original action plan established in 1984. The new action plan allows the role of biosphere reserves to be reviewed in accordance with the emphasis on sustainable development brought about by UNCED, held in Rio in 1992.

Besides a new action plan the Seville conference articulated a new vision for biosphere reserves. Because of its relevance to the theme of this book, the Seville vision is presented here in full.

### The Seville Vision

The vision for biosphere reserves into the twenty-first century which was established at the Seville conference is as follows (UNESCO 1995b).

> What future does the world face as we move towards the 21st century? Current trends in population growth and distribution, globalisation of the economy and the effects of trade patterns on rural areas, the erosion of cultural distinctiveness, increased demand for energy and resources, centralisation and the increasing difficulty of access to information, and uneven spread

# BIOSPHERE RESERVES 217

of technological innovations – all these paint a sobering picture of environment and development prospects in the near future.

The UNCED process laid out the alternative of working towards sustainable development, incorporating care of the environment and greater social equity, including respect for rural communities and their accumulated wisdom. Agenda 21, the Conventions on Biodiversity, Climate Change, and Desertification, and others, show the way forward at the international level.

But the global community also needs working examples that encapsulate the ideas of UNCED for promoting both conservation *and* sustainable development. These examples can only work if they express all the social, cultural, spiritual and economic needs of society, and they are also based on sound science.

Biosphere reserves offer such models. Rather than becoming islands in an increasingly impoverished and chaotic world, they can become theatres for reconciling people *and* nature, they can bring knowledge of the past to the needs of the future, they can demonstrate how to overcome the problems of the sectoral nature of our institutions.

Thus biosphere reserves are poised to take a new role. Not only will they be a means for the people who live and work within and around them to retain a balanced relationship with the natural world, they will also contribute to the needs of society as a whole by showing us a way to a more sustainable future. This is the heart of our vision for biosphere reserves in the 21st century.

Ten key directions were identified by the conference, which are the foundations of the new Seville strategy.

1. Strengthen the contribution which biosphere reserves make to the implementation of international agreements promoting conservation and sustainable development, and especially to the Convention on Biological Diversity, Climate Change, Desertification and others.
2. Develop biosphere reserves in a wide variety of environmental, economic and cultural situations, from largely undisturbed regions to the peripheries of great cities. There is a particular potential, and need, to apply the biosphere reserve concept in the coastal and marine environment.
3. Strengthen the emerging regional/inter-regional and thematic networks of biosphere reserves as components within the world network of biosphere reserves.
4. Reinforce scientific research, monitoring, training and formal education in biosphere reserves since conservation and sustainable use in these areas require a sound base in the natural and social sciences, and the human sciences. This need is particularly acute in countries where biosphere reserves lack human and financial resources.
5. Ensure that all zones of biosphere reserves contribute to conservation, sustainable development and scientific understanding.
6. Extend the transition area to embrace large areas suitable for ecosystem management and use the biosphere reserve to explore and demonstrate approaches to sustainable development at the regional scale. More attention should be given to the transition area. In short, this view of a biosphere reserve is wider than that of a protected area.
7. Reflect more fully the human dimensions of biosphere reserves. Connections should be made between cultural and biological diversity. Traditional knowledge and genetic resources should be conserved and

their role in sustainable development should be recognised and encouraged.
8. Promote the management of each biosphere reserve essentially as a 'pact' between the local community and society as a whole. Management should be open, evolving and adaptive. Such an approach will help ensure that reserves and their local communities are better placed to respond to external political, economic and social pressures.
9. Bring together all interest groups and sectors in a partnership approach to biosphere reserves both at site and network levels. Information should flow freely among all concerned.
10. Invest in the future. Biosphere reserves should be used to further our understanding of humanity's relationship with the natural world, through programmes of public awareness, information and education, based on a long-term, inter-generational perspective.

## BIOSPHERE RESERVE CASE STUDIES

In this section we describe a number of biosphere reserves from Asia, Central America, Europe, Africa and Australia, focusing on the ways in which the various reserves have linked community development with nature conservation, highlighting social research input into the process.

### Xilingol Biosphere Reserve*

Xilingol Biosphere Reserve is interesting because any success in conserving the remarkable biodiversity of the grasslands depends totally on developing a successful partnership between the reserve management and the local herders and their institutions. If the biosphere reserve model can achieve a measure of success in achieving sustainable development in the grasslands of Inner Mongolia, then this would indicate it may be a successful policy instrument for elsewhere in China and beyond.

### *China's biosphere reserves*

China is one of the world's 'mega-biodiverse' countries, with over 10% of the world's flowering plants, and about 10% of its mammals, birds, reptiles and amphibians (Brattz 1992). Since the first nature reserve was nominated in China in 1956, the area of territory protected and the number of reserves have grown steadily, apart from during the period of the cultural revolution when great damage was done to China's natural environment (Li & Zhao 1989). A rapid expansion over the last decade or so resulted in the declara-

---

* The report on this case study is based on research by the authors and their colleagues in the Chinese Academy of Sciences, the China MAB Committee and the Xilingol Biosphere Reserve Management, reported in part in Thwaites *et al.* (1995) and Li *et al.* (1995). See also Chapter 6 for further information on research carried out within the reserve.

tion of 763 nature reserves by the end of 1993, covering 661 000 km$^2$, or about 6.8% of the country, up from only 34 reserves covering 1265 km$^2$ in 1978 (Han & Guo 1995).

However, despite this enormous growth of the network of protected areas, there are numerous problems facing the reserves. China contains 20% of the world's population and has suffered dramatic habitat loss, funds are severely limited for protected area management, and there is a critical lack of well trained staff (Brattz 1992). A study of the results of a questionnaire sent out to reserves across China concluded 'that institutional arrangements are generally lacking, funding is critically short, staffing is undermined by poor conditions, protection measures are limited by several factors, and the integration of conservation and the development of the local economy is critical to the success of the programme' (Han & Guo 1995).

The biosphere reserve programme has been seen as a particularly useful concept in China, with its focus on the combination of conservation and development into one programme. China nominated its first MAB biosphere reserve in 1979, and now has 10 nature reserves which have joined in the international network of biosphere reserves designated by UNESCO. Following this international model, China has set up its own Biosphere Reserve Network (CBRN), with 45 designated biosphere reserves as at 1995, introducing the concept and objectives of the international network into their own design and management (Xu 1995).

*Characteristics of Xilingol reserve*

The Xilingol Biosphere Reserve is situated in Inner Mongolia Autonomous Region, about 600 km north of Beijing (Map 2), and was established as China's first grasslands nature reserve in 1985 to protect the biodiversity of a typical steppe ecosystem, and to develop models of sustainable grassland resource use for improved wellbeing of the local people (Li *et al.* 1995). The reserve joined the international MAB biosphere reserve network in 1987.

The Xilingol Biosphere Reserve is situated in the southeast part of the Mongolian Plateau, and covers an area of 10 786 km$^2$, corresponding to the catchment of an inland flowing river, the Xilingol River. The area experiences a cool temperate semi-arid climate, and covers a range of landscape types from 950 to 1500 metres above sea level. The resultant variety of local climatic features, soil and vegetation types contains a great ecosystem and species diversity. Research in the reserve has identified 655 species of flowering plants, 71 species of mosses, more than 10 species of large fungi, 76 species of birds, 33 species of mammals, 33 grasshoppers and other abundant insects (Li *et al.* 1995).

Control over various aspects of the reserve's management and development is a very complex issue, with numerous departments at different levels of government holding certain responsibilities. Xilingol Biosphere Reserve is situated within the Xilingol League of the Inner Mongolia Autonomous Region, and covers an area roughly overlapping that of the Xilinhot City (equivalent to a 'banner' or county). Xilinhot city itself sits at the centre of Xilingol Biosphere Reserve, and contains a population of roughly 100 000 people. The reserve also covers all or part of four livestock farms, Baiinxile, Maodeng, Beilike and Baiinkulun. All of these levels of control, as well as grazing households and individual herders, make decisions which may impact upon the management of the grassland resource within the biosphere reserve.

The main functional area of the reserve, including the areas nominated as core areas and buffer zones, is situated in the Baiinxile Livestock Farm, which covers an area of 3680 km$^2$ and has a population of about 11 000 (Li et al. 1995).

*Rapid rural appraisal research*

Within the reserve the Inner Mongolian Grasslands Ecosystem Research Station (IMGERS) of the Chinese Academy of Science undertakes fundamental and applied studies of the grasslands ecosystem. Recently a collaborative research programme has been established between IMGERS and the Johnstone Centre of Parks, Recreation and Heritage at Charles Sturt University, Australia, to investigate social and policy issues. In 1994 a rapid rural appraisal (RRA) was undertaken to establish an information base to assist reserve management planning, suggest policy reform and establish a longitudinal research programme (Thwaites et al. 1995).

The RRA fieldwork included a variety of information-collecting techniques, including group interviews, semistructured interviews, site visits and observations, as well as gathering information from a range of secondary data sources. Time was spent getting a feel for the geography of the area, population, relationship of the people to the land, land use activities etc. Observation played a key role in gathering much of this information. Site visits were undertaken to a number of landscape features, demonstration and trial plots, development sites and villages, and usually included interviews with relevant people or groups associated with that site.

These visits and interviews offered a better understanding of how the biosphere reserve and the Baiinxile Farm operate: their relationship; a range of activities and issues related to the planning and management of the reserve and the farm; implementation of these plans; and land management issues associated with the grazing practices of the herders.

Interviews were held with a number of key informants including decision makers. These were officers associated with the UNESCO and Chinese Biosphere Reserve Network, management of the Xilingol Biosphere Reserve and management of the Baiinxile Farm, local government officers

BIOSPHERE RESERVES                                                221

in Xilingol League (the regional government) related to land and environmental management and tourism, and scientists from the Chinese Academy of Sciences and University of Inner Mongolia, who have carried out research over the years in the fields of ecology and economics on the Xilingol grasslands in Baiinxile Farm. Interviews were also held with a number of local farmers and herders in different places. Sometimes these were associated with site visits to trial plots and demonstration sites, and sometimes they were carried out on an opportunistic basis on the grasslands and in villages. Group interviews were also held with groups of mostly men in the streets of villages, and with some local women employed by IMGERS as research assistants.

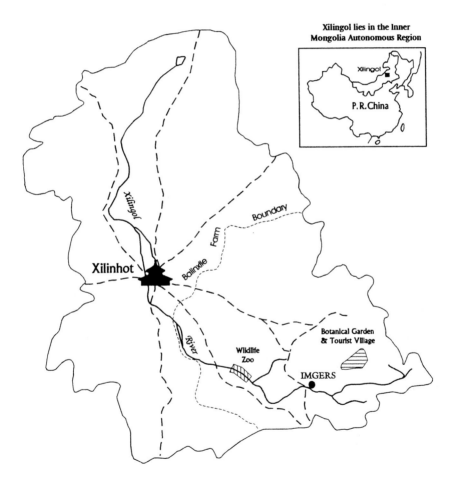

Map 2   Xilingol Biosphere Reserve

Through this exploratory rapid rural appraisal, a background picture of the Xilingol Biosphere Reserve and the Baiinxile Farm was able to be built up. Issues of importance to biosphere reserve management, to the maintenance of the biodiversity of the grassland, and to the development of sustainable alternative economic activities were able to be identified. Although more detailed work needs to be undertaken to develop a true picture of the problems and solutions (see Chapter 6 on rural development), key points have been identified.

*Grazing*

The issue of primary concern to conservation is grazing management. The vast biodiversity of the Xilingol grasslands is in the huge range of grasses and flowering plants, and the animal species these support. Animal husbandry is the mainstay of the economy of the Xilingol grasslands, and the local community considers the grasslands to be their basic resource.

**Plate 10.1** Herder with his cattle in Xilingol Biosphere Reserve. In Xilingol, the major biodiversity is found in the grassland species of plants. Grazing, the primary economic activity of the local people, is presently impacting on this biodiversity. (Photo: R. Thwaites)

*Land degradation* Poor grazing management is already threatening this industry. An increase of stock numbers and changes to grazing regimes have resulted in serious grassland degradation and desertification. Land degradation was universally recognised in the area, by farm managers, herders and villagers interviewed during the field work. It was typically associated with the amount of grass available, described in comments such as 'when I was young there was more grass', or 'the grass used to be this high, now you can see there is hardly any grass'.

*Grazing patterns*  The grasslands of the Mongolian Plateau have evolved over many thousands of years with migratory grazing of wild ungulates and traditionally nomadic grazing of domesticated stock, including sheep, goats, cattle, horses and camels. These grazing patterns have greatly changed since the coming to power of the communist party in 1949. Nomadic groups were settled into villages, to form collectives and production teams, forming a permanent stationary presence on the grasslands, particularly around sources of permanent water. The policy of economic development of the grasslands, and a greatly increased income to herders, has led to a major change in their lifestyle. Permanent houses have been constructed as well as enclosed stables for animal wintering, and the nomadic grazing pattern has virtually disappeared from Inner Mongolia.

Research undertaken at IMGERS has shown that maximum biodiversity is achieved with light grazing pressure, in the form of high intensity grazing for a short period, followed by a period free from grazing. Total species diversity decreases with either a decrease or an increase in grazing pressure. With greatly increased grazing pressure and no periods free from grazing, both the species diversity and the quantity of grass are reduced so greatly that the grassland can be seriously degraded.

Since the establishment of the Baiinxile Livestock Farm in 1950, some areas around the permanent villages and water supplies have become seriously degraded. Within the biosphere reserve some of these areas have been fenced off for rehabilitation, while less accessible areas away from the villages have remained in relatively good condition.

The research suggests that traditional nomadic patterns, under which the biodiversity has evolved over thousands of years, would be the preferred grazing pattern from the point of view of biodiversity. It is therefore desirable to establish a grazing management pattern which mimics the effect of the nomadic grazing, while recognising the social realities and desires of the local people for a better life – rotational grazing on fenced plots.

*Stock numbers*  As well as recognising the occurrence of widespread land degradation, it was also a widely held view from both herders and farm management that there is no control on the number of stock that individuals can have. It is up to individual herders to decide on the size of their stock. There may be several reasons for this increase in livestock numbers. Longworth (1993) identifies a strong link between the increasing population and increasing livestock numbers. A number of respondents in Xilingol Biosphere Reserve related the increase in livestock numbers to the Mongolian people's cultural association of wealth to the number of animals owned. Government policies and reforms have also encouraged herder households to expand their livestock numbers, and increased the impact of those animals on the grasslands.

*Impact of reforms – double contract system* In the pastoral regions of China, reforms were introduced in 1979 to invigorate the economy by offering management responsibility to individual households, and linking remuneration to output. The introduction of the 'double contract household production responsibility system' involved the transfer of animals from collective or village ownership to household ownership, to be raised under contract, and also the distribution of state or collective land under contract to households, providing them with the right to manage that land. The ability to own animals privately and reap the rewards is a strong incentive to increase the size of the flock.

The other contract of the 'double contract' system is the right to manage allocated land. However, only a small part of the available land appears to be allocated in such a way, and certainly all the better land around good water supply is grazed in common, as is the summer grazing grounds. The ownership of some pastures is also under dispute between subvillages, as well as between the Baiinxile Farm and a neighbouring farm. There is also some uncertainty as to the responsibility required of the household regarding management of the land, and as such allocations are only for limited periods of time and reassessed at intervals, there is no strong relationship built up in the minds of the herders between the ecological state of their grazing land and the future of their flocks.

*Economic growth – increased demand* The massive economic growth in most sectors in China, particularly in the urban economies, has raised demand for agricultural products. The increased demand for a variety of animal products provides a strong incentive for herders to increase the size of their flocks and herds. Without corresponding changes in policy as mentioned above in relation to long-term responsibility for land management, there is little incentive to conserve the grassland resource.

The increase of stock numbers and changes to grazing regimes have resulted in serious grassland degradation and desertification. Such degradation reduces the carrying capacity of the land and restrains the development of the animal husbandry industry, putting great economic pressure on the increasing population.

*Reserve management*

The management of the reserve itself is the responsibility of the Biosphere Reserve Management Bureau, under the Department of Urban and Rural Construction and Environmental Protection of the Inner Mongolia Autonomous Region. However, Xilingol League and Xilinhot City, as different levels of local government, both hold responsibility for certain factors in the area, from road construction to urban planning. Responsibility for land management rests mostly with the farms who use the land. The Biosphere Reserve Management Bureau is left with little true power for decision making, and must convince the farms to achieve any real change in the area.

*Relationship with Baiinxile Farm*  The relationship between the farms and the Reserve Management Bureau is then of prime importance in meeting the objectives of the biosphere reserve. The reserve Management Bureau has focused all its energies so far on establishing a working relationship with the state-run Baiinxile Farm, and no attempt has been made to introduce the concept of the biosphere reserve or its objectives into the other farms.

*Core areas*  To meet the requirements of the biosphere reserve programme, four core areas have been nominated for conservation purposes. These are small fenced areas, where all human activity, including hay cutting, grazing and gathering of plants, is banned.

*Demonstration farm*  The Xilingol Biosphere Reserve has also established a demonstration farm, to develop a range of alternative experimental farming techniques, establish sustainable development industries based on the grassland other than wool and meat production, and provide an economic return to supplement the funds of the reserve. The reserve believes that the demonstration farm can act as an important education resource for the local community and for visitors on alternative uses of the grassland and nature conservation.

*Ecotourism*  Recognising the need to diversify the local economy, both the Xilingol Biosphere Reserve and Baiinxile Livestock Farm recognise the potential of the area for tourism, and strongly support the development of ecotourism opportunities. The tourism industry in China is growing very rapidly, with an increase in demand from both overseas visitors and the expanding middle classes of China. Local officials and people involved in developing this tourism potential have identified the exotic culture of the Mongolian grasslands, and the natural beauty of the region as prime attractions.

Four sites are currently identified as ecotourism opportunities and three projects are being developed. A very diverse area containing forests, woods, steppe, meadows and marshes in the sand dune country is being developed as a botanical garden of native species. This will include the construction of facilities such as greenhouse, herbarium, display hall and offices. A tourist holiday village has been established near this botanical garden, beside a lake. This village provides tourists with the opportunity to live in *yurts*, experience some of the local Mongolian herder culture, and the natural features of the lake and the grasslands, particularly in the botanical gardens. It has also been proposed to establish a wildlife zoo in an area covering some sand dune country and hill steppe, and containing the Xilingol River. Wildlife native to the area, but now very rare or non-existent, will be reintroduced, such as the Mongolian horse, gazelle, fox and deer. The fourth site is the Great Wall of Ghengis Khan, which stretches across the grasslands at the southern end of the biosphere reserve. This site is just a short drive from the tourist village, and no specific development is proposed here at present.

All the land being developed for these projects is owned by the Baiinxile Livestock Farm, and the tourism developments are cooperative projects between the farm and the Biosphere Reserve Management Bureau. As with all other aspects of the development of the biosphere reserve, this cooperation is crucial for the continued development of the objectives of the biosphere reserve.

### Boucle de Baoule Biosphere Reserve[*]

The Baoule Loop Biosphere Reserve in Mali is currently undergoing a rehabilitation programme, implemented by the Malian government, United Nations Development Programme and the World Bank. The programme is aimed at improving the management of the reserve through 'participatory contracts' with around 50 000 local people living in and around the reserve. Some of the issues being addressed by the programme include the presence and expansion of villages, uncontrolled nomadic stock grazing and illegal hunting of native species by poachers. These pressures have contributed to the gradual degradation of the reserve.

---

[*] This case study is based on Traore (1995) and a report by Niare (1995) at the UNESCO International Conference on Biosphere Reserves held in Seville, Spain, 20–25 March 1995.

The programme initially revised the boundaries of the reserve in consultation with local communities. The reserve now separates villages and stock corridors from areas set aside for rigorous conservation.

The second step of the rehabilitation programme involves extensive community development and involvement in the management of the reserve. Measures such as supporting improvements to health services for the people, promoting cash-generating activities and assisting the implementation of training projects for local people have been undertaken to improve the socioeconomic conditions of local communities. Practical training workshops, such as teaching local hunters how to carry out a wildlife inventory, have been held to promote community interest in the sustainable use of natural resources in the area.

The management of the reserve and surrounding areas is to be improved by implementing a reserve management plan which involves extensive community input, along with the introduction of grazing management plans for the areas surrounding the reserve. Following production of the reserve management plan, the managers of the reserve hope to involve the local community in management activities such as reserve protection, inventory work and development activities. The establishment of Natural Resource Management Village Committees involves the community in the preservation of the area's natural resources within and outside the reserve.

### Rhoen Biosphere Reserve*

The Rhoen Biosphere Reserve is significant in that it is an excellent example of integrated regional planning involving the biosphere reserve model. Conservation of the characteristics of the reserve relies on the continuation of human use of the reserve's resources. Such use requires extensive planning to ensure that sustainable levels of resource utilisation are maintained, and that suitable methods are engaged to preserve the existing landscape characteristics. The management bodies responsible for the reserve have developed a framework management plan for the reserve to manage activities within the area.

*Characteristics of the reserve*

Rhoen Biosphere Reserve is located in Germany, and spans the border of three Länder (federal states) – Bavaria, Hessen and Thuringia. This impacts

---

* A paper describing the management of the Rhoen Biosphere Reserve was presented at the UNESCO Conference on Biosphere Reserves held in Seville in March 1995. This paper (Pokorny 1995), along with a summary of the framework management plan (State Ministry of Development and Environmental Affairs of Bavaria (SMDEA)(nd), form the basis for this case study.

on the management of the reserve, as each Land has separate administration arrangements. The reserve was adopted by UNESCO in 1991 to protect both the cultural landscape and the natural faunal and floral species remaining in the area.

The core zone occupies just 2% of the total reserve area of 185 000 hectares. The core zone consists mainly of relatively natural forest areas which are to be allowed to regenerate without further direct impact from human activities. The buffer zone covers nearly 40% of the area of the reserve. The zone comprises areas which have been consistently utilised for agricultural purposes and currently support a rich variety of floral and faunal species. The buffer zone also encompasses many pockets of near-natural forests. The zone of influence occupies the largest percentage of the reserve, around 60% of the total area. Industrial areas, towns and villages occur within this zone, along with the forestry and agricultural sites of most significance to the economy. Most of the 111 000 people living in the Rhoen reside within the zone of influence.

The history of land use within the reserve has moulded the present characteristics of the region. Traditional land use methods need to be maintained to preserve the current mix of species in the area along with the unique character of the cultural landscape. In order to maintain such land use practices, extensive communication and education programmes with the residents are essential.

*Management zones*

The reserve has been divided into management zones which correspond to the core, buffer and transition zones. For the purposes of managing activities within the reserve, the zones have been renamed, and the buffer zone divided into two development levels. The core zone was named the conservation zone, to reflect the land use within the area and assist public understanding of the aims of the reserve. The buffer zone was divided into maintenance zones A and B. Maintenance zone A allows for minimal access, to promote the preservation of floral and faunal species within the area. Maintenance zone B allows more access than zone A, but limits development and provides a buffer zone between the high conservation values of zone A and the conservation zone and the permissive development nature of the transition zone.

The transition zone was renamed the development zone, again to reflect the land use of the area. The purpose of the development zone is to act as a model region, demonstrating ways people can utilise the biosphere without causing unacceptable environmental changes. The reserve managers promote economic activities undertaken in the development zone which are carried out in an environmentally sustainable manner whilst providing for the economic and social requirements of the population.

*Economic activity within the reserve*

The Rhoen is traditionally a highly productive agricultural area of Germany, which has also been used for forestry and other industries. The area is still used for a variety of economic purposes. The maintenance of the multifaceted nature of economic activity within the reserve is seen to be essential for the continuation of sustainable economic activity in the region. The framework management plan identified objectives for the development of the key economic activities within the reserve, through consultation with the community and an assessment of the preservation requirements of the natural and cultural resources.

*Agriculture* The maintenance of traditional agricultural land use practices is essential to maintain the Rhoen's biodiversity and ensure the preservation of the area's unique character. Along with forestry, agriculture is one of the most important economic activities within the Rhoen. The managers of the reserve are working with the local community to introduce a multifaceted agricultural management strategy to be implemented across the reserve. The strategy will offer economic alternatives for farmers and ways to boost productivity without having detrimental effects on the natural features of the area or the cultural landscape.

*Forestry* The reserve encompasses a large number of pockets of natural or near-natural forests. Some of the more diverse forests were included in the core zone areas of the reserve, but many occur in the buffer and influence zones. The framework management plan identifies the 'guiding framework' for forestry within the reserve as 'the requirement to manage the whole of the forest in accordance with near-natural or naturalistic maintenance methods' (SMDEA nd: 16).

*Tourism and recreation* Tourism is the major growth industry within the reserve. It relies heavily on the maintenance of the cultural landscape for which the reserve was adopted, which in turn relies on sustainable land use. Therefore, local communities are encouraged to cater for sustainable levels of tourism and recreation within the reserve. The tourism and recreation visitation to the area provides an opportunity for the reserve management to work with private enterprise to monitor the quality of the visitor experience. The management bodies are encouraging the expansion of the network of restaurants and guest houses in the region and the sale of locally produced goods at markets or local shops. Activities will be encouraged which focus on the features of the reserve, involve education regarding the cultural and natural value of the area and promote an understanding of the biosphere reserve concept.

*Population and economy* The economic viability of the reserve is as essential as the preservation of its natural and cultural characteristics. To achieve both goals, the conservation of the area's natural and cultural features must be brought into line with the economic development of the Rhoen. Programmes are being implemented to encourage development and land use practices which allow for economically viable and sustainable use of the area's resources.

*Transport systems* One of the significant developments within the reserve is the rationalisation of the transport system, to improve service and minimise impact on the landscape. Existing cycleways, roads and railway lines are being upgraded to allow for extensive tourist use, and to encourage use of public transport options among the local community.

*Other land uses* Along with the major economic activities outlined above, many other land use issues are addressed by the framework management plan in an attempt to monitor their development within the reserve. These include: water management (including ground water, surface water and waste water); the introduction of wind power plants; the use of regenerative energy sources (including woodchips for heating and the generation of biogas from organic material); the continuation of basalt extraction within the reserve; the promotion of hunting within the reserve (primarily to control the deer population); recreational use of the reserve; and the development of former military installations.

*Sustainable development solutions*

A range of initiatives have been developed by the reserve managers to maximise sustainable development within the region. Figure 10.3 illustrates the process of achieving the objectives for protection, maintenance and development of the reserve through sustainable development. The promotion of suitable types of economic activity within the reserve is an integral part of the conservation of the reserve's natural resources. One of the programmes to assist local development is the introduction of a regional logo, which is used by local farmers and other businesses to indicate the environmentally friendly production of goods and services within the region. An ecological industrial area will also be established which will incorporate industrial plants where local agricultural products can be processed. This will reduce transport times and costs and minimise pollution from processing activities.

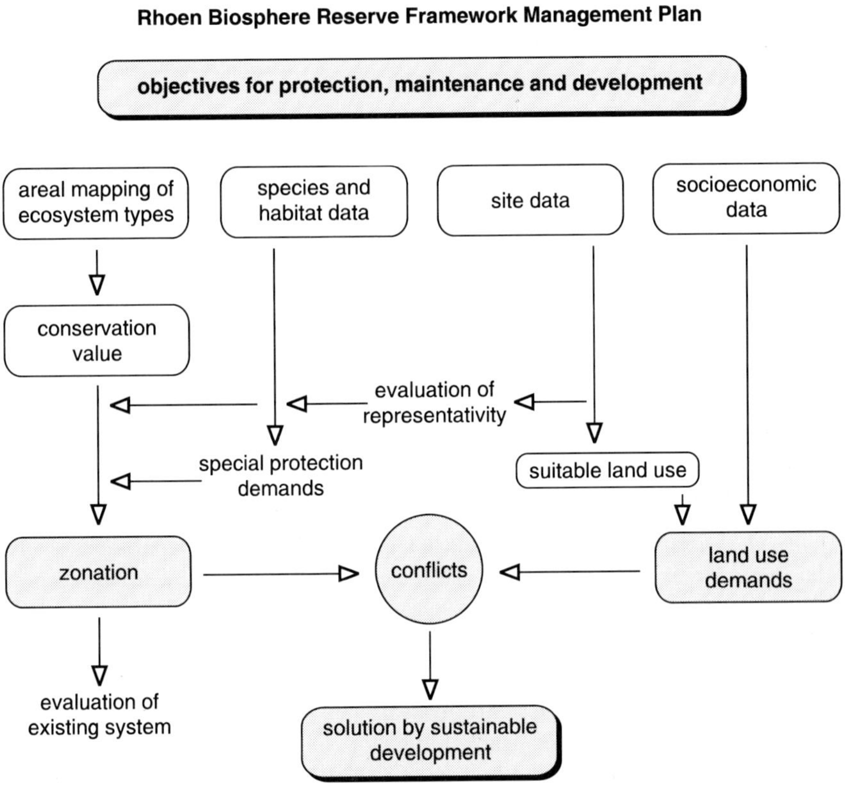

# BIOSPHERE RESERVES

**Rhoen Biosphere Reserve Framework Management Plan**

**objectives for protection, maintenance and development**

areal mapping of ecosystem types → conservation value

species and habitat data

site data

socioeconomic data

evaluation of representativity

special protection demands

suitable land use

zonation → conflicts ← land use demands

evaluation of existing system

solution by sustainable development

Figure 10.3  Achieving sustainable development through the framework management plan (after Pokorny 1995)

*Philosophy and management of the biosphere reserve*

The management structure for the reserve reveals the importance of community integration into the process. The small area of core zone in the reserve leaves nearly 98% of the reserve to be managed while being utilised by the community. The framework management plan was developed with extensive community involvement to provide the basis for the protection, maintenance and development of the reserve. During the production of the plan, the community was involved through discussion with the district councils, the municipalities which occur within the reserve, local authorities and NGOs, along with a number of informative events, distribution of interim reports, integration of over 150 submissions and suggestions to the plan, questionnaires, workshops and seminars.

The manager of the Rhoen Biosphere Reserve stated that 'the success of the Biosphere Reserve depends on the identification of the local people with this project, their commitment and cooperation' (Pokorny 1995: 5). Such community involvement in the production of the framework management plan was essential as the plan itself has no legislative status, and relies on the voluntary adoption of its principles by local communities and planning authorities. The sense of ownership which arises from community involvement in the planning process has assisted the adoption process.

Already some of the programmes proposed within the plan have been adopted by local community groups, including increasing cooperation between local farmers and restaurants. The willingness of the community to adopt the proposed management strategies reveals the success of the community involvement process.

It is important that many of the measures outlined in the plan are implemented by private enterprise, which can span the boundaries of the official planning authorities. Local communities and private enterprise can operate relatively independently of the Länder boundaries which separate the official management authorities for the reserve. The reliance on private enterprise, local communities and regional planning authorities introduces the requirement for a coordinating body to oversee the implementation of the framework management plan throughout the reserve, and to evaluate the effectiveness of the strategies and programmes set up. The framework management plan proposes the establishment of the 'Rhoen Agency' to act as a support body for the individual communities and associations working within the reserve. The agency would also assist in coordinating the activities of the individual groups.

### Island of Hiiumaa (West-Estonian Archipelago Biosphere Reserve)*

The West-Estonian Archipelago Biosphere Reserve, which encompasses the Island of Hiiumaa, has undergone extensive socioeconomic changes in recent years. In an attempt to coordinate the community responses to these changes with the objectives of a biosphere reserve, a 'Development Concept' has been introduced to manage the reserve.

The managers of the reserve are attempting to implement sustainable resource use principles in the developing market economy system on the island. In order to maximise exposure of the goals of the 'Development Concept', the reserve managers are engaging popular media such as TV, radio and newspapers to convey their message.

---

\* This case study is based on a report presented by Kokovkin (1995) to the UNESCO International Conference on Biosphere Reserves held in Seville, Spain, 20–25 March 1995.

The integration of the community into preservation of the island's resources is being achieved through increasing local participation in resource management and improving education and awareness of the ecological characteristics of the island. This involves field days where aspects of community interaction with the natural environment are discussed (such as 'fishermen and seals' and 'forestry days'), participation of members of the community in work carried out in natural areas, training administrators and private enterprise operators in natural resource management issues, working with educational institutions on co-projects and holding lectures and seminars for the community.

Along with maintaining close contact with the island community, the managers of the reserve maximise their involvement in the biosphere reserve network and international cooperation opportunities. The dynamic nature of the socioeconomic status of the community of the island will continue to have a major impact on management of the biosphere reserve, and on the use of the island's natural resources.

### Mapimi Biosphere Reserve*

Mapimi Biosphere Reserve in Mexico is an excellent example of the integration of community development and conservation in protected areas.

#### Characteristics of the reserve

The reserve is located in northwest Mexico, and spans the borders of three states – Durango, Chihuahua and Coahuila. Unlike the Rhoen Biosphere Reserve, a single management body – the Institute of Ecology – is responsible for the whole reserve. The reserve is representative of the desert *matorral* which extends across a vast area of the centre of the American continent. The reserve is characterised by a wide variety of vegetation associations and structures which results in a variety of faunal species. The reserve occurs within the Bolson de Mapimi province, and occupies 100 000 hectares of the Chihuahuan Desert. The reserve was established in 1977 as a representative example of the central American desert ecosystem and to protect the endemic great desert tortoise.

Historically the area has been used for cattle grazing, an industry which remains the primary economic activity in the region. The adoption of the

---

* Features of the management programme of Mapimi Biosphere Reserve have been recorded by a number of researchers and this case study has drawn on many of these. A description of the establishment of the reserve (Halffter *et al.* 1977), records of the role of local participation in the conservation and development of the reserve (Halffter 1981 and 1984) and an account of the social aspects of the biosphere reserve (Kaus 1993) form the basis for this case study (see Chapter 4 for further information on research in the reserve).

area as a biosphere reserve has had minimal impact on the grazing rights of the local community. However, the reserve status has limited the amount of hunting in the area, particularly of the great desert tortoise (*Gopherus flavomarginatus*). Prior to adoption of the reserve, the tortoise was extensively hunted and in danger of extinction.

The desert environment is harsh and generally supports only around 200 people. The land tenure of the reserve includes freehold land (mainly used for cattle ranching) and usufruct rights over government land. The Institute of Ecology has exclusive rights over a small area (20 hectares) of land in the centre of the reserve where it has established a field study station.

*Biosphere reserve activities*

Core areas have been delineated within the reserve with the assistance of the local cattle graziers. The buffer zone within the reserve is primarily used by cattle ranchers and graziers, while many of the villages around the reserve are located within the zone of influence or transition zone. Cattle grazing is traditionally a noble occupation throughout northern Mexico, and remains the primary economic activity of the community. The reserve area sustains a low density of cattle due to the vegetation type and abundance. Some of the research being carried out on the reserve involves determining alternatives to improve the carrying capacity of the area for cattle production. Most cattle production is for export to the USA.

Other land uses in the reserve include harvesting plants, such as the candelilla (*Euphorbia antisyphilitica*) for wax which is used in the production of candles. Salt is extracted from two ephemeral saline lakes within the zone of influence of the reserve. Small plots of corn, beans, squash, melons and other vegetables are raised for household purposes and for feeding livestock. Agricultural production on a larger scale is generally not carried out, partly due to the people's preference for cattle ranching.

Apart from cattle production, the most prolific activity within the reserve is scientific research. Since the establishment of the reserve, research has been carried out regarding the ecology and many other aspects of the area. The establishment of a field study station in the centre of the reserve allows for extensive research projects to be carried out on site. One of the major benefits of the research is the integration of the local community with the research groups. Local people are employed to perform support work for the researchers, and local graduates are recruited to be involved with the scientific work. Along with the ecology of the reserve, research work has focused on the effects of cattle grazing, ways to increase agricultural production without increasing the amount of land worked, and other studies of the practicalities of integrating human use of the land with conservation of the natural resources.

Social research has also been a major focus within the reserve. Much of the research has been based on analysing the relationship between the local community and the reserve's researcher–managers. One of the major issues revealed by the research is that the local inhabitants see their relationship with the researchers, which includes informal agreements on certain activities within the reserve, as a serious social contract. The contract binds the residents into certain actions, such as eliminating the use of poison bait to kill coyotes (*Canis lantrans*) and not hunting the great desert tortoise, in a much stronger way than would official laws and regulations. A second issue which arose was the significance assigned by the local community to cattle ranching. Cattle ranching is seen as a noble, traditional way of life, and is one which is unlikely to stop in the near future. The population of the reserve is dynamic due to the nature of the cattle grazing industry. Generally, families which run cattle on the reserve operate a household in a town or village nearby in addition to a house within the reserve. This allows the family to remain in the town for the majority of the year while one or two members of the family return to the reserve to monitor the cattle. The family members living on the reserve often change throughout the year, and are generally men or older boys.

*Philosophy and management of the biosphere reserve*

Management of the Mapimi Biosphere Reserve has involved local participation from its inception. The location and boundaries of the reserve were determined by a group of local residents, lead by a cattle rancher. Through this method, the managers of the reserve (the Institute of Ecology) could be sure that the local community was not threatened by the imposition of a conservation programme in an inappropriate area. Once the location of the reserve was established, a legally constituted, community-based management body was set up. The management body consists of representatives of the local cattle ranchers and graziers, small landowners, the National Council of Science and Technology, the Institute of Ecology and the MAB programme of Mexico. The management body improves links between various groups within the reserve's local inhabitants, and between the community and government authorities.

Along with the preservation of a representative example of the desert ecosystem and the maintenance of a population of the great desert tortoise, a major objective of the reserve is the improvement of the socioeconomic status of the community both within the reserve and surrounding it. The Institute of Ecology, which coordinates the research programmes carried out on the reserve, also has extensive administration responsibilities for the reserve. This provides for effective administration of the reserve, as information regarding the ecology and socioeconomic features of the reserve is freely available to the responsible authority. Possibly as a result of its dual role, the institute carries out a range of research programmes focusing on maximising economic gains from the region while minimising environmental costs.

The enforcement of laws and policing regulations within the reserve is strictly limited by the management authority in favour of self-monitoring and regulation being carried out by the local inhabitants. In this way, the managers encourage the maintenance of the social contract which has developed with the reserve's residents. The arrangement has been quite successful to date, particularly in minimising poaching by people from outside the reserve. However, such success may be due to the local inhabitants seeing such poachers as a threat to their livelihood, rather than the strong association they have established with the researcher–managers. The success of the programme may also rely on the absence of cost to the local inhabitants of cooperating with the researcher–managers.

The status of the community-based management body of the reserve has been strengthened by the support of various state and federal government officials and departments. The Mapimi Biosphere Reserve's community-based management structure, along with the combination of conservation and improved agricultural production within the reserve, act as an example of successful integration of community development and nature conservation through biosphere reserves.

### Mae Sa-Kog Ma Biosphere Reserve*

The Mae Sa-Kog Ma Biosphere Reserve in northern Thailand exemplifies the problems associated with many protected areas throughout the mountainous mainland of southeast Asia; that is, striking a balance between the requirements of the local human population and protection of the biodiversity of the mountain ecosystem. The Mae Sa-Kog Ma Reserve is unique in terms of its highly valued economic, cultural and natural resources, as well as its proximity to the major urban centre of Chiang Mai.

The reserve was established in 1977 and covers 14 200 hectares, supporting 20 000 residents and an increasing number of tourists. The reserve overlays the catchment of the Sa River, and encompasses a large portion of the Doi Suthep-Pui National Park. The area is one of the most highly populated mountain regions in Thailand and consists of many minority ethnic groups. The forests of the area are extremely diverse, sustaining 1959 plant species and 1265 animal species. Many species present within the reserve are endemic to the area, and a large number are considered rare or endangered.

---

* This case study is based on Rerkasem & Rerkasem (1995) and WCMC/UNESCO (1994).

The principal use of the reserve by humans has been shifting agriculture and harvesting forest products. From the mid-1970s a programme was initiated within the Doi Suthep-Pui National Park to steer agricultural production towards market supply. This increased the area of land used for rice paddies and introduced shifting crops on the steeper slopes. Since then, the shifting crops have mostly been replaced by cash crops. This has decreased the area of land required by the residents for economically sustainable production. However, the advantages of the transition have been offset by increasing population in the area, primarily due to immigration of lowland Thais hoping to farm the cash crops.

Apart from agriculture, the main activities currently within the reserve are tourism and scientific research. Mae Sa-Kog Ma Biosphere Reserve is unusual in Thailand for its proximity to a major town, Chiang Mai, and the high standard of road access to the reserve. Research carried out within the reserve has focused mainly on establishing sustainable agricultural practices, assessing the impact of introduced species, and studying the effectiveness of existing methods for utilising the area's natural resources. Tourism within the area is increasing each year. The tourism industry is based on the natural and cultural features of the reserve, including villages, temples and waterfalls. The tourists include both Thais and international visitors. The tourism industry has brought development into the reserve in the form of commercial operations and an improved road network. Some villages within the reserve have developed tourism-based industries to a point where agricultural production is secondary to tourist facilities.

Management of the reserve is complicated by a number of factors. First, there is no legislation to support the establishment of the biosphere reserve. This has resulted in the attempt to promote biosphere reserve policies within the national park, which covers a major portion of the reserve. Secondly, the majority of the reserve is owned by the central government, with the remaining areas privately owned. The area owned by the government is further divided, as a range of government departments have administrative responsibilities for various parts of the reserve and different aspects of its development.

Management of the national park has recently benefited from a new emphasis on consultation with the local population. The development of plans for the management of the national park have involved determining community opinion regarding proposed development projects prior to establishing policies. In recent years a number of developments have been postponed due to community protest, including a proposed cable car from Chiang Mai to a cultural site, and the construction of a training centre in the heart of the national park. Future directions in the management of the reserve will be towards more extensive community involvement in management decisions and the possible division of the national park into areas for specific development objectives and management strategies. Areas of high tourist pressure will be managed according to policies, independent of the rest of the park, which deal with the specific issues relevant to such areas. The remaining areas of the park, where the principles of coordinated conservation and sustainable development are more relevant, will be managed in accordance with the objectives of a biosphere reserve.

**Bookmark Biosphere Reserve***

Bookmark biosphere reserve is especially important for the fact that emphasis is being placed heavily on the *social context* that surrounds the pursuit of the reserve's conservation goals. Work at Bookmark to develop ecologically sustainable development activities to be conducted in the biosphere reserve is especially relevant, and has involved a great deal of local involvement and participation. The next stage, the actual implementation of such activities if they are found to be economically and scientifically feasible (including acceptable in conservation terms), is yet to be tested, but the present situation offers hope that the important work so far carried out may continue.

*Characteristics of the reserve*

Bookmark is located in South Australia, adjacent to the junction of the borders between New South Wales, Victoria and South Australia. The area is semi-arid to arid, with low and erratic rainfall, and the original vegetation type is mallee, a form of low, stunted eucalypt tree assemblage. Part of the Murray River, Australia's longest river and the most important in the Murray Darling Basin (which covers most of Australia's southeast and supports much of Australia's agricultural production), is included in the reserve. The wetlands of the river and its anabranches in the reserve are recognised by international wetland conventions, and support many flora and fauna species.

*Environmental problems* The irrigated dry landscape has suffered many problems common in Australia, including soil salinity increases, loss of perennial plant cover, soil erosion, water quality decline, pollution and nutrient runoff, pest species, and reducing productivity, increasing costs, and the imposition of economic costs on people downstream. In addition, people in the area are affected (almost always adversely) by resource man-

---

* This case study is based on research by the authors and from secondary sources, most importantly Parker *et al.* (1995).

agement decisions made upstream (a classic example of an economic externality, where economic costs are not contained within a system and incurred by the person causing them but passed on to others). One of the major environmental changes associated with the river has been the modified flooding regime, where the natural patterns of flooding and drying have been completely changed by the impoundment of water for irrigation. The salinity problem is one of the most serious of environmental concerns, and is caused by rising water tables as a result of irrigation and forest clearing. It is in this degraded and highly stressed environment that the biosphere reserve is operating.

**Plate 10.2**   Pelicans on anabranch of the Murray River, in Bookmark Biosphere Reserve. Conservation of biodiversity is one of the objectives, or functions of BRs, and many of the wetlands on the Murray floodplains within the reserve are registered under the RAMSAR treaty. (Photo: R. Thwaites)

*History of conservation*   The federal government purchased three pastoral leases in the area and gave them to the South Australian Government to be managed as a protected area – Dangalli Conservation Park. This park was designated a biosphere reserve in 1975, although the biosphere reserve concept in its entirety was not embraced, and many of the requisites for the area to be considered a biosphere reserve were not met.

Dangalli Conservation Park formed the original core area of the biosphere reserve. In 1992 the Australian Nature Conservation Agency reviewed the biosphere reserve programme in Australia, and identified problems including lack of cooperation between levels of government, lack of resources, and especially lack of knowledge about community management of land for conservation and ecologically sustainable development. The Fitzgerald River Biosphere Reserve (in the southwest of Australia) was held up as a good example of this integration, and the potential for attempting the same in the ecologically important mallee was recognised. In 1992 the purchase of Calperum Station, a collection of pastoral leases, added significantly to the potential of the biosphere reserve to contribute to conservation of biodiversity and the process of achieving ecological sustainability. Calperum was purchased under an innovative public/private arrangement involving the government and the Chicago Zoological Society, and has been transferred to the Australian Nature Conservation Agency under the provisions that it must be used for public recreation, public education (by being a model for the ecologically sustainable use of resources), for the protection of wetlands and bird species, and in accordance with the biosphere reserve action plan. The overall area (including Dangalli National Park, Calperum Station, Murray River National Park, and Chowilla Regional Reserve) was nominated as Bookmark Biosphere Reserve in 1993. The local community in this area, known as the Riverland, was receptive to the idea of the biosphere reserve and already had some experience in management issues in the area.

*Social and production characteristics* The majority of the area in and around the biosphere reserve was devoted to pastoral activity, with pastures irrigated from the Murray river system. In addition to grazing, there is some fruit and other crop production (also utilising irrigation) and some dryland cropping. The local people and the local economy rest on the irrigated industries, which are also the cause of the majority of the environmental problems facing the area. Local people appear to place a high value on the landscape, not only because of its economic importance to them, but also for recreational and cultural reliance on the river system for many activities, including fishing, hunting, camping and boating and the aesthetic values of the area. The grazing of sheep, which has been the dominant activity in most of the reserve, has been assessed as unsustainable due to high stocking rates. Grazing of sheep on Calperum has been discontinued to alleviate this, pending consultation and research to ascertain whether sheep grazing can be an appropriate, economically viable and sustainable use of the land.

*Biosphere reserve activities*
The body that arose from the Chowilla process was formed and recognised legislatively as the Murraylands Conservation Trust. This trust has assumed

authority for all reserves now part of the Bookmark Biosphere Reserve, and is responsible for maintaining and protecting land and resources in the reserve, adapting the biosphere reserve action plan of UNESCO in a regionally relevant manner and in cooperation with the relevant statutory agencies, and continuing consultation with the community. In 1993, two community meetings were held to address the future of the reserve. One meeting addressed the community's role in the management of the land in the reserve, a role shared with the statutory authorities. The other meeting addressed the role of determining what forms of sustainable land and resource use could be supported by the reserve. This meeting, with the help of a professional facilitator, uncovered a vision for the area that included a healthier river and wetlands, a more diverse and vigorous economy, employment for local people, better land productivity and extensive natural areas for recreation. In order to look at some of these issues, a working group was formed for examining ecologically sustainable development (ESD) activities as a means of moving towards these goals. This working group defined ESD, and through examining literature on ESD decided on the characteristics that ESD activities needed to have (such as inter- and intragenerational equity, compatibility with degraded land and so on). The working group also took into account the factors affecting communities beyond their own when assessing potential activities.

*Development activities being investigated*

The ESD working group for the Calperum section of the biosphere reserve decided on 11 categories of potential ESD activities that required investigation. The criteria used for the selection of these activities included conservation considerations (judged to be paramount), the retention of aesthetic and recreational values of the land and activities which deal with excess animal populations (native and feral), assist with fire management, provide local employment and investment potential, restore the natural landscape and promote water quality, native biota and physical land components. The committee also stressed the fact that, to understand fully the potential of ESD approaches for the Riverlands, there needs to be an ongoing process. The committee also became involved in informal education within the community and made more technical investigations into aspects of the particular ESD activities. The committee also stressed the need for ongoing monitoring of environmental and other aspects of the performance of these activities.

One element of the Bookmark approach is that business plans for these activities have or will be prepared, and the operations will then be offered to local people for investment in the form of franchises from the biosphere reserve's holdings.

The categories of potential ESD that were developed are as follows.

*Feral animal control*  Control of feral animals through harvesting, through mustering and delivery to local processors, and through hunting. The main species involved are goats and rabbits, both of which occur in large numbers across the reserve. The functions that can be fulfilled through this activity include the control of these pest species, the provision of recreational hunting opportunities, and the provision of new industries and local employment. A full financial analysis has been undertaken by business consultants (who donated their time) who recommended a business plan they considered viable.

*The utilisation of native wildlife*  Some native wildlife species, especially kangaroo and emu, are unnaturally high in the area due to the provision of water for stock and the availability of feed. As well as controlling the impact that these species have on adjacent lands, harvest will allow the reduction of these species to levels within the carrying capacity of the landscape.

*Grazing of domestic stock species*  The grazing of domestic stock may be possible if numbers are limited to sustainable levels. This type of activity will also depend on the development of economically sound local markets for producing value-added products, in order to insulate the industries against fluctuations in international commodity markets. These industries will protect existing jobs as well as create new opportunities.

*Honey production*  The production of honey from the native flowering plants is to be considered, but will depend on the scientific assessment of the effect of European bees on the native flora and fauna. Regular monitoring of the industry was also recommended as part of a permit-granting process.

*Harvest of plant materials from cropping and natural growth*  Plant collection for food, ornamental, medicinal and other uses is being considered which will incorporate a programme of monitoring and review of these plants and their landscape setting. One particularly promising form of utilisation is the production of industrial solvents from essential oils found in many mallee species. One of the dominant mallee species in the area produces large amounts of oil. The presence of cleared areas within the mallee may allow plantations for these purposes to be established, allowing these activities to proceed without compromising the ecological integrity of the more intact areas. In areas of high salinity there is potential for use of halophytic plants to provide cover for the soil, and provide table oil (from pressing) and stock fodder (after pressing). The Saudi Arabian government has been developing these techniques using plants of the same genera as some of those occurring in the reserve. In addition to these activities, native flower, live shrub and tree cropping and collection for a variety of purposes, such as bush food and provision of plants for restoration, will also be considered. Cropping of more conventional crops on the upland (non-floodplain) areas may also be possible with evaluation of environmental factors and the employment of sustainable farming principles.

*Paid hunting* Hunting is also proposed, within the limitations of existing statutes, linked with habitat recovery programmes and population monitoring of target species. The ESD working group feels that any paid hunting venture should be coordinated and managed with recognised hunting organisations who have conservation and education objectives as part of their mission.

*Ecotourism* Ecotourism, based around the principles of education, reference to the biosphere reserve programme, low levels of environmental impact, and local involvement and benefit, is presently undergoing detailed investigation by the business consultants. Ecotourism activities for the area may include educational and special interest tours, for birdwatching, photography and so on. Any ecotourism developments may be able to make use of facilities on the Calperum portion of the reserve, which include shearers' quarters, meeting rooms and kitchens (in three separate locations).

*Aquaculture* Aquaculture is also being considered as one of the ESD activities. A number of fields on the floodplain had been banked and laser levelled prior to the purchase of Calperum station. These fields will be used as pondages for aquaculture, with the production of native fish and/or brine shrimp being considered. Native fish may find markets as food as well as for reintroduction to other areas where possible. Native fish stocks locally have been decimated and made locally extinct by the presence of the pest fish European carp. The effects of pondages on salinity and other environmental issues (such as possible disease problems) need to be assessed and taken into account when developing the aquaculture proposals.

*Mining* Mining in the reserve was considered acceptable by the ESD working group with the provisos that the flood plain is not directly threatened, that enough income is made to allow full restoration of mine sites and that full environmental impact assessments are carried out. The opportunities offered by mine restoration works to allow experimentation on methods of restoration of natural vegetation, and the possibilities for such sites to be used for other ESD activities may also be important aspects of this activity.

*Future*

The history of local community involvement in the management of natural resources began in 1990, when the Murray Darling Basin Commission (a body set up to manage the huge catchment of the Murray and Darling rivers) Ministerial Council agreed to involve the community in the development of a resource management plan for Chowilla Regional Reserve. The resultant establishment of Bookmark Biosphere Reserve and development of local community ESD groups has laid the foundations for a very exciting journey. The process to date has allowed local people to gain valuable experience in the planning and management process. It also has given government agencies and staff experience in being involved in genuine participatory planning as opposed to the directive role normally played in a consultative process. Bookmark is capturing international interest as a model process for linking local development to conservation through protected areas.

CONCLUSION

Considering the major theme of this book – how to link local level development more effectively with conservation through protected areas – and reviewing a host of case studies reveals at least one certainty: the requirement for both genuine local community participation (bottom up) and genuine national and regional government commitment (top down). Unless this partnership is in place, no longlasting integration of conservation and development appears possible. The tragedy is that this happens so rarely.

Many internationally funded integrated conservation/development projects have, in the past, worked through central government agencies without including local communities as integral parts of the process. Alternatively, in recent times, NGOs have put much work into securing local community participation but have not always been able to secure genuine central government commitment to an ongoing programme. It therefore seems appropriate for us to look at particular public policies that specifically provide for this crucial local community/central government linkage. If these particular policy instruments can be made to work, such models can inform sustainable development initiatives internationally (with appropriate modifications for different contexts).

Governments are more likely to put into place relevant enabling legislation and provide administrative structures (and even funding) if there are international policy models they can use, and obtain the political benefits of being seen to implement such models. If programmes that have local community support are developed under the auspices of these policies, they are more likely to gain the genuine commitment of government and its bureaucracies. The biosphere reserve model is such a policy. It has international stature, deliberately fosters a partnership between government and local community and aims for ecologically sustainable development, and successful examples are starting to emerge.

# A Final Note

There have been a number of themes running through this book. We have emphasised the need for local participation both in an ethical and in a practical sense. We have emphasised also the uses of a social science approach to understanding the ways in which development is a conservation issue and conservation is a development issue. This has been introduced in the first section of the book.

We have also introduced the reader to social science applications, in terms of analysis of issues (rural development, indigenous people and tourism), management approaches (local resource management and biosphere reserves) and, more generally, as processes of change which are underpinned by social science approaches.

Of these themes, there has been one which we have kept coming back to: that of conservation being the same as development (that is, participatory development). The whole search for sustainability should be linked to the quest for societies, cultures and groups of people having more control over their futures and the decisions which shape their lives.

Such an approach overturns dominant ideas about development, and puts in place a belief system, an ethical model and a management practice which is in keeping with a humanitarian approach to social and conservation issues.

However, whilst we have emphasised this, it is important to avoid the perils of dogmatism and all that it implies. While we have emphasised a local level approach and explored examples within a social science framework, the social science approach also highlights the relative nature of social, economic and political life. There are no final 'truths' as they are known in the sciences. Rather, there are any number of complex layers of understanding which local/participatory approaches must try to unravel in their search for sustainability. There is little that is right or wrong – there are, however, cultural, social, economic and political contexts within which people's definitions and perceptions of right and wrong differ. We are faced then with a balancing of differing perceptions, with alterations to notions of good and bad, right and wrong, and with the necessity to change, where appropriate, these as well as the institutional arrangements which support them.

An integrated approach must avoid dogmatism. The experience of such approaches and our understanding of social science approaches tell us that there are many pathways forward, and many side roads to be taken before a more direct journey can be undertaken. The importance for an integrated approach, we believe, is that the journey be undertaken in a flexible way but with goals in mind that are the result of local level participation, consultation, needs and wants. In other words, the journey is one which represents a variety of concerns and approaches, not just dogmatic assertions of the inherent right or wrong of various views.

We believe that protected area management has an important role to play. Existing protected area systems represent institutionalised systems of conservation (albeit with limitations) and consequently provide a vehicle for global conservation. New protected areas, new models of management and the critical appraisal of biosphere reserves provide potential for the further use of participatory approaches to conservation and development, where creative relationships between local people, protected area management, non-government organisations and/or state and international agencies can be facilitated.

What this book has also addressed is the reality – the reality of forging creative relationships such as those mentioned above, of integrating vision with implementation, of balancing the pragmatic with the ideal. This has led into an exploration of alternative approaches to the search for sustainability. Included in this is an examination of some of the ways in which the rights of local people can be, and have been, integrated with institutional arrangements which have characteristics often associated with more top-down traditions.

The experience of the cases within this book, and social science insights in general, have much to offer the search for sustainability. The task is to ensure that the potential is translated into action and reality.

# References

Adventure Travel Society (1994) Newsletter, Winter edition. Englewood, Colorado.
Allcock, A., Jones, B., Land, S. & Grant, J. (1994) *National ecotourism strategy*. Commonwealth Department of Tourism, Canberra.
Anderson, C. (1992) Identifying groups for negotiation about land. In R. Hill (ed.) *Cross cultural management of natural and cultural heritage*. Cairns College of Technical and Further Education, Cairns, Australia.
Asch, M. (1984) *Home and native land – Aboriginal rights and the Canadian constitution*. Methuen, Toronto.
Bailey, K. (1987) *Methods of social research*. Free Press, New York.
Barnes, J., Burgess, J. & Pearce, D. (1992) Wildlife tourism. In T. Swanson & E. Barbier (eds) *Economics for the wilds: wildlife, wildlands, diversity and development*. Earthscan, London.
Batisse, M. (1986) Developing and focusing the biosphere reserve concept. *Nature & Resources* **XXII** (3): 2–11.
Batisse, M. (1993) The silver jubilee of MAB and its revival. *Environmental Conservation* **20**: 107–12.
Bennett, T. & Blundell, V. (1995) Introduction: first peoples. *Cultural Studies* **9**: 1–10.
Birckhead, J., De Lacy, T. & Smith, L. (eds) (1993) *Aboriginal involvement in parks and protected areas*. Aboriginal Studies Press, Canberra.
Birckhead, J. & Wallis, A. (1994) *An evaluation of the Cairns college of TAFE 'Community Ranger' program Final Report*. The Johnstone Centre of Parks, Recreation and Heritage, Charles Sturt University, Albury.
Birckhead, J., Klomp, N. & Roberts, A. (1996) Protocols for the participation of Aboriginal communities in research and monitoring. In M. Bomford & J. Caughley (eds) *Sustainable use of wildlife by Aboriginal peoples and Torres Strait Islanders*. Australian Government Publishing Service, Canberra.
Brandon, K. (1993) Basic steps toward encouraging local participation in nature tourism projects. In K. Lindberg & D. Hawkins (eds) *Ecotourism: a guide for planners and managers*. The Ecotourism Society, North Bennington, Vermont.

Brattz, S. (1992) *Conserving biological diversity: a strategy for protected areas in the Asia-Pacific region.* World Bank technical paper No. 193, World Bank, Washington.

Brennan, F. (1994) *Sharing the country.* Penguin Books, Melbourne.

Brettell, C. (1993) Introduction: Fieldwork text and audience. In C. Brettell (ed.) *When they read what we write – the politics of ethnography.* Bergin and Garvey, Westport, Connecticut.

Bridgewater, P. (1992) Strengthening protected areas. In *Global biodiversity strategy: guidelines for action to save, study, and use earth's biotic wealth sustainably and equitably.* WRI, IUCN, UNEP, FAO and UNESCO, Washington.

Burger, J. (1990) *The Gaia atlas of first peoples – a future for the Indigenous world.* Penguin Books, London.

Burkey, S. (1993) *People first: a guide to self-reliant, participatory rural development.* Zed Books, London.

Byers, A. (1994) *Understanding behavioural motivations: a key to integrating conservation and development.* To Fifth International Symposium on Society and Resource Management, 7–10 June 1994, Fort Collins, Colorado State University.

Campbell, A. (1989) Landcare in Australia – an overview. *Australian Journal of Soil and Water Conservation* 2: 4.

Campbell, A. (1994) *Landcare: communities shaping the land and the future.* Allen & Unwin, Sydney.

Campbell, J., Palit, S. & Roy, S. (1994) *Putting research partnerships to work: the joint forest management research network in India.* To Fifth International Symposium on Society and Resource Management, 7–10 June 1994, Fort Collins, Colorado State University.

Ceballos-Lascurain, H. (1993) Ecotourism as a worldwide phenomenon. In K. Lindberg & D. Hawkins (eds) *Ecotourism: a guide for planners and managers.* The Ecotourism Society, North Bennington, Vermont.

Cernea, M. (ed.) (1991) *Putting people first: sociological variables in rural development.* Oxford University Press, London.

Chambers, R. (1986) *Rural development: putting the last first.* Longman, London.

Chambers, R. (1991) Shortcut and participatory methods for gaining social information for projects. In Cernea, M. (ed.) *Putting people first: sociological variables in rural development.* Oxford University Press, Washington.

Clairborne, Liz, and Art Ortenberg Foundation with Michael Wright (1993) *The view from Airlie.* Liz Clairborne and Art Ortenberg Foundation, Washington.

Clifford, J. (1986) Introduction: partial truths. In J. Clifford & G. Marcus (eds) *Writing culture: the poetics and politics of ethnography.* University of California Press, Berkeley.

Cordell, J. (1993) Who owns the land? Indigenous involvement in Australia's protected areas. In E. Kemf (ed.) *The law of the mother: protecting indigenous people in protected areas.* Sierra Club Books, San Francisco.

Craig, D. (1990) Social impact assessment: politically oriented approaches and applications. *Environmental Impact Assessment Review,* **10**: 37–54.

Curtis, A. & De Lacy, T. (1995) Landcare evaluation in Australia: towards an effective partnership between agencies, community groups and researchers. *Journal of Soil and Water Conservation* **50**: 15–20.

Curtis, A. & De Lacy, T. (1996a) Landcare in Australia: does it make a difference. *Journal of Environmental Management,* in press.

Curtis, A. & De Lacy, T. (1996b) Landcare in Australia: beyond the expert farmer. *Journal of Agriculture and Human Values,* in press.

Curtis, A., Birckhead, J. & De Lacy, T. (1994) Community participation in landcare policy in Australia: the Victorian experience with regional landcare plans. *Society and Natural Resources,* **8**: 415–30

Curtis, A., Tracey, P. & De Lacy, T. (1993) *Landcare in Victoria: getting the job done.* Johnstone Centre, Albury, Australia.

Dahlan, H. (1990) In what way can culture serve tourism? *Borneo Review* **1**: 129–48.

Dasmann, R. (1976) Life-styles and nature conservation. *Oryx* **13**: 281–6.

Davidson, A. (1993) *Endangered peoples.* Sierra Club Books, San Francisco.

Davis, S. (1993) *Indigenous views of land and the environment.* World Bank Discussion Papers, No. 188, The World Bank, Washington.

De Lacy, T. (1994) The Uluru/Kakadu Model – Anangu Tjukurrpa. 50,000 Years of Aboriginal Law and Land Management Changing the Concept of National Parks in Australia. *Society and Natural Resources* **7**: 479–98

De Lacy, T. & Lawson, B. (1995) The Uluru/Kakadu model: joint management of Aboriginal-owned national parks in Australia. In S. Stevens (ed.) *Conservation through cultural survival: national parks, protected areas, and indigenous people.* Island Press, Washington

DCE (Department of Conservation and Environment) (1992) *Victoria's decade of landcare plan.* DCE, Melbourne.

de Vaus, D. (1991) *Surveys in social research.* Allen and Unwin, Sydney.

Dickman, S. (1989) *Tourism: an introductory text.* Edward Arnold, Melbourne.

Dillman, D. (1978) *Mail and telephone surveys: the total design method.* Wiley/Interscience, New York.

Driver, B., Dustin, D., Baltic, T., Elsner, G. & Peterson, G. (eds) (1996) *Nature and the human spirit: toward an expanded land management ethic.* Venture, State College, Pennsylvania.

Dube, S. (1988) *Modernisation and development: the search for alternative paradigms.* Zed Books, London.

Farley, R. & Toyne, P. (1989) A national land management programme. *Australian Journal of Soil and Water Conservation*, **11**: 6–9.

Federation of Nature and National Parks of Europe (1993) *Loving them to death?: sustainable tourism in Europe's nature and national parks.* Federation of Nature and National Parks of Europe, Grafenau.

Ferraro, G., Trevathan, W. & Levi, J. (1992) *Doing cultural anthropology: anthropology, an applied perspective.* West Publishing: Minneapolis.

Filion, F., Foley, J. & Jacquemot, A. (1994) The economics of global ecotourism. In M. Munasinghe & J. McNeely (eds) *Protected area economics and policy: linking conservation and sustainable development.* World Bank and IUCN, Washington.

Finsterbusch, K., Llewellyn, L. & Wolf, C. (eds) (1983) *Social impact assessment methods.* Sage, Beverly Hills.

Freeman, M. & Carbyn, L. (eds) (1988) *Traditional knowledge and renewable resource management in northern regions.* Occasional publication, Boreal Institute for Northern Studies, The University of Alberta, No. 23, Edmonton.

Freire, P. (1972) *Pedagogy of the oppressed.* Penguin Books, Melbourne.

Furze, B. (forthcoming) *Gandhian approaches to local level conservation and development: toward contemporary relevance.*

Furze, B. & Stafford, C. (1994) *Society and change: a sociological introduction to contemporary Australia.* Macmillan Education Australia, Melbourne.

Giddens, A. (1989) *Sociology.* Polity Press, Cambridge.

Giddens, A. (1992) *Human societies: a reader.* Polity Press, Cambridge.

Gionjo, F., Bosco-Nizege, J. & Wallace, G. (1994) *A study of visitor management in the world's national parks and protected areas.* Colorado State University, The Ecotourism Society, IUCN & WCMC.

Glaser, M. & Marcus, R. (1994) *Community based nature tourism in Belize: a way to protect the forests.* To Fifth International Symposium of Society and Resource Management, Fort Collins, Colorado State University.

Greaves, T. (1995) Cultural rights and ethnography. *General Anthropology*, **1**: 1–6.

Guba, E. & Lincoln, Y. (1989) *Fourth generation evaluation.* Sage, London.

Halffter, G. (1981) Local participation in conservation and development. *Ambio* **10**: 93–6.

Halffter, G. (1984) Conservation, development and local participation. In F. Di Castri, F. Baker & M. Hadley (eds) *Ecology in Practice.* UNESCO & Tycooly International Publishing Limited, Dublin.

Halffter, G., Barbault, R. & Celecia, J. (1977) Mapimi and Michilia, two biosphere reserves in Latin America. *Nature and Resources* **13**: 18–20.

Ham, C. & Hill, M. (1985) *The policy process in the modern capitalist state.* Wheatsheaf, London.

Han, N. & Guo, Z. (1995) An investigation into the current situation of China's nature reserves. *China's Biosphere Reserves*, Special English Issue. China MAB, Beijing.

Harriss, J. (ed.) (1984) *Rural development: theories of peasant economy and agrarian change.* Hutchinson, London.

Hartley, R., Riches, J. & Davis, J. (1992) A systems approach for landcare. In *Proceedings Volume 1 of the 7th International Soil Conservation Organisation Conference – People protecting their land*, 27–30 September 1992, Sydney.

Hawke, R. (1989) *Our country our future: statement on the environment by the Prime Minister of Australia.* Australian Government Publishing Service, Canberra.

Healy, R. (1992) *The role of tourism in sustainable development.* To the IVth World Parks Congress on National Parks and Protected Areas, February 1992, Caracas, Venezuela.

Hobart, M. (ed.) (1993) *An anthropological critique of development – the growth of ignorance.* Routledge, London.

Internet (1995a) *What is world heritage?* (nd: retrieved via Netscape 1995, July 6) http://kaos.erin.gov.au/land/conservation/wha/what_wha.html

Internet (1995b) *Center for world Indigenous studies: advancing cooperation and consent between nations.* (nd: retrieved via Netscape 1995, July 11) http://www.halcyon.com/FWDP/cwisinfo.html

Internet (1995c) Parkipuny, M. *The Indigenous peoples rights question in Africa* (1993: retrieved via Netscape 1995, July 11)ftp://ftp.halcyon.com/pub/ FWDP/Africa/parkipny.txt

Internet (1995d) Peeters, Y. *On the discrimination of the Rehoboth Basters: an Indigenous people in the Republic of Nambia* (1993: retrieved via Netscape 1995, July 24)ftp://ftp.halcyon.com/pub/FWDP/Africa/rehoboth.txt

Interorganisational Committee on Guidelines and Principles for Social Impact Assessment (1994) *Guidelines and principles for social impact assessment.* US Department of Commerce, National Oceanic and Atmospheric Administration, National Marine Fisheries Service, Washington.

IUCN (1993) *Parks for life: report of the IVth World Congress on National Parks and Protected Areas.* IUCN, Gland.

IUCN (1994) *Guidelines for protected area management categories.* CNPPA and WCMC. IUCN, Gland.

Jacobs, M. (1991) *The green economy: environment, sustainable development and the politics of the future.* Pluto Press, London.

Jonas, W. (1991) *Consultation with Aboriginal people about Aboriginal heritage.* Australian Heritage Commission, Canberra.

Kaus, A. (1992) *Common ground: ranchers and researchers in the Mapimi Biosphere Reserve.* A dissertation submitted in partial satisfaction of the requirements for the degree of Doctor in Philosophy in Anthropology. University of California, Riverside.

Kaus, A. (1993) Social realities of environmental ideologies: a case study of the Mapimi Biosphere Reserve. *Culture and Agriculture* **45**: 29–34.

Keesing, R. (1981) *Cultural anthropology: a contemporary perspective.* Holt, Rinehart & Winston, New York.

Kemf, E. (1993) *The Law of the Mother: Protecting Indigenous Peoples in Protected Areas.* Sierra Club Books, San Francisco.

Kiss, A. (ed.) (1990) *Living with wildlife: wildlife resource management with local participation in Africa.* World Bank technical paper number 130: Africa technical department series, The World Bank, Washington.

Kleymeyer, C. (1993) *Cultural traditions and community based conservation.* Paper prepared for The Liz Claiborne Art Ortenberg Foundation Community Based Conservation Workshop, 18–22 October 1993, Airlie, Virginia.

Knudtson, P. & Suzuki, D. (1992) *Wisdom of the elders.* Allen & Unwin, Sydney.

Kokovkin, T. (1995) *People and biosphere reserves: Island of Hiiumaa (West Estonian Archipelago Biosphere Reserve): building a self-supported developing community.* To UNESCO International Conference on Biosphere Reserves, 20–25 March 1995, Seville, Spain.

Kovall, R. (1994) Dance to the music of time. *New Scientist Supplement*, 26 November.

Kruijer, G. (1987) *Development through liberation: third world problems and solutions.* Macmillan Education, Basingstoke.

Kuper, L. (1981) *Genocide.* Penguin Books, Suffolk.

Lewis, H. (1993) Traditional ecological knowledge: some definitions. In N. Williams & G. Baines (eds) *Traditional ecological knowledge: wisdom for sustainable development.* Centre for Resource and Environmental Studies, Australian National University, Canberra.

Li, B. (1993) *Protected area strategy for east-Asia.* IUCN, Gland.

Li, W. & Zhao, X. (1989) *China's nature reserves.* Foreign Languages Press, Beijing.

Li, Y., Yan, Y., Yong, S. & Chen, Z. (1995) Xilingol grassland reserve: role of scientific research in linking nature conservation with local development. *China's Biosphere Reserves*, Special English Issue. China MAB, Beijing.

Lindberg, K. & Enriquez, J. (1994) *An analysis of ecotourism's economic contribution to conservation and development in Belize, volume 2: comprehensive report.* WWF and the Ministry of Tourism and the Environment, Belize.

Lindberg, K. & Hawkins, D. (eds) (1993) *Ecotourism: a guide for planners and managers.* The Ecotourism Society, North Bennington, Vermont.

Lindberg, K. & Huber, R. (1993) Economic issues in ecotourism management. In K. Lindberg & D. Hawkins (eds) *Ecotourism: a guide for planners and managers*. The Ecotourism Society, North Bennington, Vermont.

Lindberg, K. & Johnson, R. (1994) Estimating demand for ecotourism sites in developing countries: the travel cost and contingent valuation method. *Trends*, 31.

Lockwood, M., Loomis, J. & De Lacy, T. (1993) A contingent valuation survey and benefit–cost analysis of forest preservation in East Gippsland, Australia. *Journal of Environmental Management*, 38: 233–43.

Longworth, J. (1993) *China's pastoral region: sheep and wool, minority nationalities, rangeland degradation and sustainable development*. CAB International, Wallingford.

Macnaughten, P., Grove-White, R., Jacobs, M. & Wynne, B. (1995) *Public perceptions and sustainability in Lancashire: indicators, institutions, participation*, report by the Centre for the Study of Environmental Change, Lancaster University. Lancashire County Council.

Marcus, G. & Fisher, M. (1986) *Anthropology as cultural critique: an experimental moment in the human sciences*. The University of Chicago Press, Chicago and London.

Marsden, D. (1994) Indigenous management and the management of Indigenous knowledge. In S. Wright (ed.) *Anthropology of Organizations*. Routledge, New York.

Maybury-Lewis, D. (1992) *Millennium: tribal wisdom and the modern world*. Viking Penguin, New York.

McCracken, J., Pretty, J. & Conway, G. (1988) *An introduction to rapid rural appraisal for agricultural development*. International Institute for Environment and Development, London.

McMillan, D. (ed.) (1991) *Anthropology and food policy – human dimensions of food policy in Africa and Latin America*. Southern Anthropological Society Proceedings, No. 24, University of Georgia Press, Athens and London.

McNeely, J., Miller, K., Reid, W., Mittermeier, R. & Werner, T. (1990) *Conserving the world's biological diversity*. International Union for Conservation of Nature, World Resources Institute, Conservation International, World Wildlife Fund-US, World Bank, Gland and Washington.

McNeely, J. & Pitt, D. (eds) (1985) *Culture and conservation: the human dimension in environmental planning*. Croom Helm, London.

Metcalf, S. (1993) *The Zimbabwe communal areas management programme for Indigenous resources (CAMPFIRE)*. Paper prepared for The Liz Claibourne Art Ortenberg Foundation Community Based Conservation Workshop, 18–22 October 1993, Airlie, Virginia.

Minichiello, V., Arone, R., Timewell, E. & Alexander, L. (1990) *In-depth interviewing: researching people*. Longman Cheshire, Melbourne.

Mitchell, R. & Carson, R. (1989) *Using surveys to value public goods: the contingent valuation method*. Resources for the Future, Washington.

Munasinghe, M. (1994) Economic and policy issues in natural habitats and protected areas. In M. Munasinghe and J. McNeely (eds) *Protected area economics and policy: linking conservation and sustainable development*. World Bank and IUCN, Washington.

Mues, C., Roper, H. & Ockerby, J. (1994) *Survey of landcare and land management practices*. ABARE, Canberra.

Murindagomo, F. (1990) Zimbabwe: windfall and CAMPFIRE. In *Living with wildlife: wildlife resource management with local participation in Africa*. World Bank technical paper 130. The World Bank, Washington.

Mutitjulu Community (1991) *Sharing the park: Anangu initiatives in Ayers Rock tourism*. Institute for Aboriginal Development, Alice Springs.

Nepali, R. (1992) *Conceptual model for activating users' groups in the Makalu-Barun area*. The Makalu-Barun Conservation Project working paper publication series, report 21. Department of National Parks and Wildlife Conservation, His Majesty's Government Nepal & Woodlands Mountain Institute, Kathmandu.

Nepali, R., Sangam, K., Ramble, C. & Chapagain, C. (1990) *The Makalu-Barun National Park and Conservation Area: community resource management component*. Department of National Parks and Wildlife Conservation, Nepal, Woodlands Mountain Institute Mount Everest Ecosystem Conservation Program, West Virginia.

Niare, M. (1995) *People and biosphere reserves: integrating local populations in the management of a biosphere reserve: the case of Boucle de Baoule*. To UNESCO International Conference on Biosphere Reserves, 20–25 March 1995, Seville, Spain.

Noonuccal, O. (1988) *The Rainbow Serpent*. Meanjin **47**: 376.

Nursey-Bray, M. and Wallis, A. (1994) *Aboriginal protocols in consultation* (Module 4, PKM 301, Aboriginal Land Management). Open Learning Institute, Charles Sturt University, Wagga Wagga.

Owen, J. (1993) *Program evaluation: forms and approaches*. Allen & Unwin, Sydney.

Palmer, P. (1989) *Environment, development and Indigenous knowledge systems: a participatory action research approach toward natural resource management in Costa Rica's Cocles/Kekoldi Indian Reserve – a thesis*. UMI Dissertation Services, A Bell & Howell Company, Ann Arbor, Michigan.

Parker, P. (1993) *Biosphere reserves in Australia: a strategy for the future*. The Australian Nature Conservation Agency, Canberra.

Parker, P., Lambie, B., Harper, M. & Anderson, I. (1995) *Community management of Bookmark Biosphere Reserve*. To UNESCO International Biosphere Reserves Conference, 20–25 March 1995, Seville, Spain.

Patton, M. (1982) *Practical evaluation*. Sage, Newbury Park, California.

Patton, M. (1987) *Creative evaluation* (2nd edn). Sage, Newbury Park, California.

Patton, M. (1990) *Qualitative evaluation methods*. Sage, Newbury Park, California.

Pearce, D., Markandaya, A., Barbier, E. (1989) *Blueprint for a green economy*. Earthscan, London.

Peters, W. (1994) *Sharing national park entrance fees: forging new partnerships in Madagascar*. To Fifth International Symposium on Society and Resource Management, 7–10 June 1994, Fort Collins, Colorado State University.

Pillsbury, B. (1984) Evaluation and monitoring. In W. Patridge (ed.) *Training manual in development anthropology*. American Anthropological Association, Special Publication No. 17: 42–63, Washington.

Pitt, D. (1985) Towards ethnoconservation. In J.A. McNeely & D. Pitt (eds) *Culture and conservation: the human dimension in environmental planning*. Croom Helm, London.

Poffenberger, M. (1993) *The resurgence of community forest management: case studies from eastern India*. Paper prepared for The Liz Claiborne Art Ortenberg Foundation Community Based Conservation Workshop, 18–22 October 1993, Airlie, Virginia.

Poffenberger, M., McGean, B., Khare, A. & Campbell, J. (1992a) *Field methods manual volume II: community forest economy and use patterns: participatory rural appraisal (PRA) methods in south Gujarat, India*. Prepared for the Joint Forest Management Support Program, New Delhi.

Poffenberger, M., McGean, B., Ravindranath, N. & Gadgil, M. (1992b) *Field methods manual volume I: diagnostic tools for supporting joint forest management systems*. Prepared for the Joint Forest Management Support Program, New Delhi.

Pokorny, D. (1995) *Biosphere reserve management: German biosphere reserve management for sustainable development*. To UNESCO International Conference on Biosphere Reserves, 20–25 March 1995, Seville, Spain.

Quarles van Ufford, P. (1993) Knowledge and ignorance in the practices of development policy. In M. Hobart (ed.) *An anthropological critique of development – the growth of ignorance*. Routledge, London.

Rajotte, F. & Bigay, J. (1981) *Beqa – Island of firewalkers*. Institute of Pacific Studies, University of the South Pacific, Suva, Fiji.

Rerkasem, B. & Rerkasem, K. (1995) *The Mae Sa-Kog Ma Biosphere Reserve, Thailand. South–south cooperation programme on environmentally sound socio-economic development in the humid tropics*. Working papers No. 3, UNESCO. The United Nations University, Third World Academy of Sciences, Paris.

Robertson, I. (1989) *Sociology*. Worth Publishers, New York.

Robinson, D. (1994) *Strategies for alternative tourism: the case of tourism in Sagarmatha (Everest) National Park, Nepal*. To Fifth International Symposium on Society and Natural Resource Management, Fort Collins, Colorado State University.

Robson, C. (1993) *Real world research: a resource for social scientists and practitioners–researchers*. Blackwell, Oxford.

Roling, N. (1988) *Extension science: information systems in agricultural development*. Cambridge University Press, Cambridge.

Sarin, M. (1993) *From conflict to collaboration: local institutions in joint forest management*. Joint Forest Management working paper number 14, National Support Group for Joint Forest Management, Society for Promotion of Wastelands Development, The Ford Foundation, New Delhi.

Sattler, P. (1991) *Towards a nationwide biodiversity strategy: the Queensland contribution*. Paper presented to Conservation Biology Conference for Australia and Oceania, October 1991, Brisbane, Australia.

Sayer, J. (1991) *Rainforest buffer zones: guidelines for protected area managers*. IUCN, Gland.

Scherl, L., Cassells, D. & Gilmour, D. (1994) *Pluralistic planning – creating room for community action in the management of the global environment*. To Fifth International Symposium on Society and Resource Management, 7–10 June 1994, Fort Collins, Colorado.

Scoones, I. (1995) Investigating difference: applications of wealth ranking and household survey approaches among farming households in Southern Zimbabwe. *Development and Change*, **26**: 67–88.

Shadish, W., Cook, T. & Leviton, L. (1991) *Foundations of program evaluation: theories and practice*. Sage: Newbury Park, California.

Simpson, T. (1991) International action on Aboriginal rights. *Habitat Australia* **19**: 30–31.

Smith, R. (1987) Indigenous autonomy for grassroots development. *CS Quarterly* **11**: 8–12.

Smyth, D. (1992) *The involvement of indigenous people in nature conservation*. To the IVth World Congress on National Parks and Protected Areas, February 1992, Caracas, Venezuela.

Society for Promotion of Wastelands Development (1993) *Joint Forest Management Update*. SPWD, New Delhi.

Spradley, J. (1986) *The ethnographic interview.* Holt, Rinehart & Winston, New York.
SMDEA (State Ministry of Development and Environmental Affairs of Bavaria) (nd) *Framework management plan – Rhoen Biosphere Reserve.* SMDEA, The Ministry of Development, Settlement, Agriculture, Forestry and Nature Conservation of Hesse and The Ministry of Agriculture, Nature Conservation and Environmental Affairs of Thuringia, Schweinfurt, Germany.
Thwaites, R., Batar, M., De Lacy, T., Furze, B. and Li, Y. (1995) *Sustainable development and biodiversity conservation in Xilingol Biosphere Reserve, China.* To UNESCO International Conference on Biosphere Reserves, 20–25 March 1995, Seville, Spain.
Tisdell, C. (1991) *Economics of environmental conservation.* Elsevier Science Publishers, Amsterdam.
Tjamiwa, T. (1993) Tjunguringkula Waakaripai: joint management of Uluru National Park. In J. Birckhead, T. De Lacy & L. Smith (eds) *Aboriginal involvement in parks and protected areas.* Aboriginal Studies Press, Canberra.
Traore, A. (1995) Tightening the loop holes. *UNESCO Sources* 69, May 1995, Paris.
Uluṟu-Kata Tjuṯa Board of Management & ANPWS (1991) *Uluṟu (Ayers Rock-Mount Olga) National Park plan of management.* Australian Government Publishing Service, Canberra.
UNEP (1995) *Nettlap News,* Vol. 3, May 1995, UNEP, Bangkok.
UNESCO (1995a) *Draft of the Seville Strategy.* UNESCO, Paris.
UNESCO (1995b) *Biosphere reserves: the vision from Seville for the 21st century.* UNESCO, Paris.
United States Parks Service (1993) *Management of ethnographic resources.* Cultural resource management NPS-28, US Parks Service, Washington.
Uphoff, N. (1986) *Local institutional development: an analytic sourcebook with cases.* Kumain Press, West Hartford, USA.
Valentine, P. (1993) Ecotourism and nature conservation. *Tourism Management* **April**: 107–15.
van Willigen, J. (1993) *Applied anthropology: an introduction.* Bergin & Garvey, Westport, USA.
van Willigen, J. & Dewalt, B. (1985) *Training manual in policy ethnography.* American anthropological association, Special Publication No. 19, Washington.
Wadsworth, Y. (1991) *Everyday evaluation on the run.* Action Research Issues Association, Melbourne.
Walsh, R. (1986) *Recreation economic decisions: comparing benefits and costs.* Venture, State College Pennsylvania.

WCED (World Commission on Environment and Development) (1987) *Our Common Future*. Oxford University Press, Oxford.

WCMC (1992) *Global biodiversity: status of the earth's living resources*. World Conservation Monitoring Centre, Cambridge.

WCMC/UNESCO (1994) *Biosphere reserves database– Thailand: Mae Sa-Kog Ma Reserve*. WCMC/UNESCO, Paris.

Wells, M. (1993) Neglect of biological riches: the economics of nature tourism in Nepal. *Biodiversity and Conservation* 2: 445–64.

Wells, M. & Brandon, K. with Hannah, L. (1991) *People and parks: linking protected area management with local communities*. The World Bank, Washington.

West, P. & Brechin, S. (1991) National parks, protected areas and resident peoples: a comparative assessment and integration. In P. West & S. Brechin (eds) *Resident peoples and national parks: social dilemmas and strategies in international conservation*. The University of Arizona Press, Tucson.

Western, D. (1993) Foreword: defining ecotourism. In K. Lindberg & D. Hawkins (eds) *Ecotourism: a guide for planners and managers*. The Ecotourism Society, North Bennington, Vermont.

Williams, N. & Baines, G. (1993) *Traditional ecological knowledge – wisdom for sustainable development*. Centre for Resource and Environmental Studies, Australian National University, Canberra.

Wilmer, F. (1993) *The Indigenous Voice in World Politics*. Sage, Newbury Park, California.

Woenne-Green, S., Johnston, R., Sultan, R. & Wallis, A. (1994) *Competing interests: Aboriginal participation in national parks and conservation reserves in Australia – a review*. Australian Conservation Foundation, Melbourne.

Wolf, E. (1982) *Europe and the people without history*. University of California Press, London.

Woodhill, J., Wilson, A. & McKenzie, J. (1992) Land conservation and social change: extension to community development – a necessary shift in thinking. In *Proceedings volume 1 of the 7th International Soil Conservation Organisation Conference – People protecting their land*. 27–30 September 1992, Sydney.

World Resources Institute, The World Conservation Union, United Nations Environment Programme (1992) *Global biodiversity strategy: guidelines for action to save, study, and use the earth's biotic wealth sustainably and equitably*. WRI, IUCN, UNEP, Washington.

Wright, P. (1993) *Ranomafana National Park Madagascar: rainforest conservation and economic development*. Paper prepared for The Liz Claiborne Art Ortenberg Foundation Community Based Conservation Workshop, 18–22 October 1993, Airlie, Virginia.

Wright, R. (1988) Anthropological presuppositions of indigenous advocacy. *Annual Review of Anthropology* **17**: 365–90.
Wright, S. (1994) 'Culture' in anthropology and organisational studies. In S. Wright (ed.) *Anthropology of organizations*. Routledge, London.
WTO (1992) *Yearbook of tourism statistics*. Madrid.
WTTC (1992) *The WTTC report: travel and tourism in the world economy*. Brussels.
Wynne, B., Waterton, C. & Grove-White, R. (1993) *Public perceptions and the nuclear industry in West Cumbria*. Centre for the Study of Environmental Change, Lancaster University, Lancaster.
Xu, Z. (1995) Editor's Notes. *China's Biosphere Reserves*, Special English Issue. China MAB, Beijing.
Young, M. (1992) *Sustainable investment and resource use: equity, environmental integrity and economic efficiency*. Man and the Biosphere Series Vol. 9. UNESCO and The Parthenon Publishing Group, London.

# Index

adventure travel 149
Africa
  indigenous people 127–8
  protected area entrance fees 160–1
  see also Mali; Nigeria; Tanzania; Zambia; Zimbabwe
agencies of socialisation 41
alienation 41
Allcock, A. 149
Anderson, C. 106
anthropology 33–7
Asch, M. 131
Australia
  Bookmark Biosphere Reserve xiv, 238–44
  community ranger program xiv, 89, 90–2, 99–101
  'Desert Tracks' 163–4
  Great Barrier Reef Marine Park 161
  Landcare, Victoria xiv, 68–9, 199–205, 206–7
  Mann Ranges xiv
  protected areas 25
  protection of old-growth forests xiv, 83–4
  Uluru-Kata Tjuta National Park xiv, 139–45, 156

Bailey, K. 66, 67
Baines, G. 138
Barnes, J. 152
Batisse, M. 209, 210–11, 212
Belize, ecotourism xiv, 159, 172–4
benefit-cost analysis (BCA) 78
  see also extended benefit-cost analysis
Bennett, T. 126
Bigay, J. 163

biodiversity 16–20
  conservation strategy 23–4
  and cultural diversity 133–4
  and protected areas 24–8
  threats to conservation 21–3
biological resources 19
biosphere reserves 27, 209–45
  Bookmark, Australia 238–44
  Boucle de Baoule, Mali 226–7
  buffer and transition zones 211–15
  case studies 218–44
  definition of 210–11
  function of 215–16
  Mae Sa-Kog Ma, Thailand 236–8
  Mapimi, Mexico 233–6
  Rhoen, Germany 227–32
  UNESCO Conference, Seville 216–18
  West-Estonia Archipelago 232–3
  Xilingol, China 218–26
Birckhead, J. vii, xiii–xvi, 90–2, 99–101, 105, 138
Blundell, V. 126
Bookmark Biosphere Reserve, Australia xiv, 238–44
Boucle de Baoule Biosphere Reserve, Mali xiv, 226–7
Brandon, K. 174, 176
Brattz, S. 218, 219
Brechin, S. 179
Brennan, F. 131
Bridgewater, P. 28
buffer zone, of a biosphere 211–15
Burger, J. 126, 127, 129, 130, 131, 132
Burkey, S. 9, 88, 89, 98, 111, 115–16
Byers, A. 50, 64

## INDEX

Campbell, A. 201, 202
Campbell, J. 189, 190, 191, 194, 198, 199
CAMPFIRE program, Zimbabwe xiv, 79, 156, 183–9, 206–7
Carbyn, L. 138
Carson, R. 82
case studies xiii–xv
  *see also by name*
Ceballos-Lascurain, H. 149, 152
Center for World Indigenous Studies 126
Cernea, M. 113–14, 115
Chambers, R. 56, 59, 111, 115, 116–17, 201
change
  agents of 14–15
  agrarian 112–13
  social 46
  *see also* development
China, Xilingol Biosphere Reserve xiv, 118–25, 155, 218–26
Clifford, J. 55
Cocles/Kekoldi Indian Reserve, Costa Rica xiv, 135–6
colonialism, and indigenous people 129–30
common-sense beliefs 41–2
community 8–9
  and conservation x, 3
  consultation 101–8
  and local level 9–10
  participation 11–13
  and property rights 180–1
community ranger program, Australia xiv, 89, 90–2, 99–101
conferences, international xiii
  *see also by name*
conflict 42
  resolution of 14–15
conscientisation 98
consciousness development, indigenous people 130–2
conservation 246
  biosphere reserves 215
  and community x, 3
  and indigenous people 136–45
  and social institutions 99
  strategy for 23–4
  threats to 28
  and tourism 146, 151, 175

consultation, community 101–8
  case study 106–8
  process 101–2, 105–6
  rationale 102–3
contingent valuation (CV) 82
Cordell, J. 143
Costa Rica
  Cocles/Kekoldi Indian Reserve xiv, 135–6
  national park 160
Craig, D. 60
Cross River National Park, Nigeria xiv, 214–15
cultural impact assessment 60–3
culture 42
  systems of meaning 55
  and tourism 161–4
Curtis, A. 68–9, 199, 200, 201, 203–4

Dahlan, H. 161, 162
Dasmann, R. 132
data analysis 67–8
Davidson, A. 129
Davis, S. 133–5
De Lacey, T. vii, xiii–xvi, 139–40, 200, 203–4
'Desert Tracks', Australia 163–4
development 4–6, 246
  local level 6–8
  rural 6, 111–25
  and social institutions 99
  sustainable 13
Dewalt, B. 86
Dickman, S. 147, 149, 162, 163
Dillman, D. 63, 67, 69
Dowdeswell, E. 4
Driver, B. 20
Dube, S. 97, 98
Durning, A. 132–3

East Usambara Mountains, Tanzania xiv, 213–14
ecological economics 40, 43
economics 37–40, 42–3
  environmental economic analysis 77–85
  and tourism 152–61
ecosystem diversity 16–18
ecosystem services 18

# INDEX

ecotourism 146–76
  Belize xiv, 159, 172–4
  Bookmark Biosphere Reserve 243
  case studies 164–74
  cultural effects 161–4
  definitions 146–52
  economic impacts 152–61
  and local development 174–6
  Madagascar 166–9
  Nepal xiv, 164–6, 169–71
Ecuador, Galapagos National Park 160
efficiency, economic 42
enclave tourism 154
Enriquez, J. 160–1, 172
environment, and indigenous people 132–6
environmental economic analysis 77–85
  case study 83–4
  extended benefit-cost analysis (EBCA) 78, 79–85
environmental economics 39–40, 43
environmental goods 44
Estonia, West Estonian Archipelago Biosphere Reserve xiv, 232–3
ethnocentrism 43, 51
ethnographic mapping 53
ethnography 51–5
  case study 53–4
  method 52–3
  and politics 54–5
evaluation 85–92
  case study 89, 90–2
  definition of 85–6
  and local level 88
  methods 87
  types of 86–7
exploitation 21–3, 129
extended benefit–cost analysis (EBCA) 78, 79–85
  case study 83–4

Farley, R. 200
Federation of Nature and National Parks of Europe 149
Ferraro, G. 52, 53
Filion, F. 149, 152
Finsterbusch, K. 60
first people *see* indigenous people
Fisher, M. 36

focus groups 75–7
forests
  joint management of, India xiv, 59, 189–99, 206–7
  protection of old-growth forests, Australia xiv, 83
Freeman, M. 138
Friere, P. 98
funnelling 73
Furze, B. vii, xiii–xvi, 42, 95

Galapagos National Park, Ecuador 160
genetic diversity 16–18
genocide 129
Germany, Rhoen Biosphere Reserve xiv, 227–32
Giddens, A. 32, 33, 47, 92, 93
Gionjo, F. 160, 161
Glaser, M. 172
globalisation 44
goods 44
Great Barrier Reef Marine Park, Australia 161
Greaves, T. 37
Guan Zuong 25
Guba, E. 2, 88
Guo, Z. 219

Halffter, G. 233
Ham, C. 35
Han, N. 219
harmony model of rural community life 9
Harriss, J. 111, 112
Hartley, R. 201
Hawke, R. 200
Hawkins, D. 156
Healy, R. 153, 154, 155
Hiiumaa, West-Estonia Archipelago Biosphere Reserve xiv, 232–3
Hill, M. 35
Hobart, M. 37
Huber, R. 157, 158–60

ideology 45
India, joint forest management xiv, 59, 189–99, 206–7

indigenous people 126–45
  case studies 135–6, 139–45
  and colonialism 129–30
  consciousness development 130–2
  definition 127–8
  and the environment 132–6
  protected areas and local
    development 136–45
institutions, social 46, 93–4
  building and strengthening
    94–101
  case study 99–101
  conscientisation 98
  and conservation and development
    99
  multilevel dimension of 95–7
International Convention on Biological
  Diversity 3, 24
International Labour Organisation
  (ILO) 130
Interorganisational Committee on
  Guidelines and Principles for
  Social Impact Assessment,
  USA 60–3
interpretivist research 50
interviewing 70–5
IUCN
  definition of protected area 26
  Fourth World Congress on Parks
    and Restricted Areas xiii, 25,
    28, 127, 137–8
  on protected area management
    26–7

Jacobs, M. 78
Johnson, R. 159
Jonas, W. 102

Kaldor-Hicks criterion 46
Kaus, A. 53–4, 233
Keesing, R. 33, 129
Kemf, E. 126, 127, 128, 129, 138
Kenya Wildlife Service 161
Kleymeyer, C. 132
Knudtson, P. 132
Kokovkin, T. 232
Kovall, R. 163–4
Kruijer, G. 115
Kuper, L. 129

Landcare, Australia xiv, 68–9,
  199–205, 206–7
Lawson, B. 139–40
legitimacy 45
  local level 56–7
Lewis, H. 14, 156
Li, B. 25
Li, W. 218
Li, Y. 218, 219, 220
Lincoln, Y. 88, 92
Lindberg, K. 156, 157, 158–61, 172
local knowledge 45
local level 10–11
  and community 9–10
  development 6–8
  legitimacy of 56–7
  management of resources 179–208
  see also indigenous people
Lockwood, M. 83
Lomgworth, J. 224

McCracken, J. 56, 57
McMillan, D. 37
Macnaughten, P. 75, 76
McNeely, J.A. ix–xi, xiii, 3, 16, 19
Madagascar, Ranomafana National
  Park xiv, 166–9
Mae Sa-Kog Ma Biosphere Reserve,
  Thailand xiv, 236–8
Makalu-Barun National Park and
  Conservation Area, Nepal xiv,
  12–13, 106–8
Mali, Boucle de Baoule Biosphere
  Reserve xiv, 226–7
management of protected areas see
  protected area management
Mann Ranges, Australia xiv
Mapimi Biosphere Reserve, Mexico
  xiv, 53–4, 233–6
Marcus, G. 36
Marcus, R. 172
market system 38–9, 45
Marsden, D. 127, 128
Maybury-Lewis, D. 132
Metcalf, S. 183, 184, 185, 188, 189
methodologies and techniques, social
  sciences 48–92
  environmental economic analysis
    77–85
  ethnography 51–5

# INDEX

evaluation 85–92
focus groups 75–7
interviewing 70–5
qualitative *vs.* quantitative research 49–50
rapid and participatory rural appraisal 56–60
social impact assessment 60–3
surveys and questionnaires 63–9
Mexico, Mapimi Biosphere Reserve xiv, 53–4, 233–6
Minichiello, V. 49–50, 70, 72, 73, 74, 75
Mitchell, R. 82
Mongolia, Xilingol Biosphere Reserve xiv, 118–25, 155, 218–26
Mues, C. 200
Munasinghe, M. 77, 78
Murindagomo, F. 183, 184, 186

Nepal
  ecotourism xiv, 164–6, 169–71
  Makalu-Barun National Park and Conservation Area xiv, 12–13, 106–8
  Sagarmatha National Park 169–71
Nepali, R. 12, 106–8
Niare, M. 226
Nigeria, Cross River National Park xiv, 214–15
Nursey-Bray, M. 102

optimal ignorance 56
Owen, J. 85

Palmer, P. 130, 135
Paretian efficiency 45–6
Pareto, Vilfredo 46
Parker, P. 215, 238
Parkipuny, M. 127
partial goods 44
participant observation 52–3
participation, community 11–13
  case study 12–13
participatory rural appraisal (PRA) 56–60
Patton, M. 85, 86, 87
Pearce, D. 38

Peeters, Y. 128
Peters, W. 166, 167, 169
Pillsbury, B. 85, 86, 87, 89–90
Pitt, D. 138
Poffenberger, M. 59, 189, 194, 196, 199
Pokorny, D. 227, 231, 232
policy 35–6
population growth 21
positivist research 50
potential Paretian improvement 46
poverty, rural 115–18
power 46
  decentralisation of 97
property rights 79, 180–1
  and tourism 155–6
protected area management x, 13, 247
  CAMPFIRE, Zimbabwe 183–9, 206–7
  case studies 183–207
  institutional linkages 182–3
  joint forest management, India 189–99, 206–7
  Landcare, Australia 199–205, 206–7
  local resource management 179–208
  property rights 180–1
  purposes of 26–7
  and rural development 111–12, 118–25
protected areas 3
  and biodiversity conservation 24–6
  classification of 26–8
  and economics 40
  and indigenous people 136–45
  threats to 28
  tourism revenues 159–61
  *see also* biosphere reserves
public goods 44

qualitative *vs.* quantitative research 49–50
Quarles van Ufford, P. 88
questionnaires 63–9

Rajotte, F. 163
Ranomafana National Park, Madagascar xiv, 166–9

rapid ethnographic assessment procedures (REAP), USA 53
rapid rural appraisal (RRA) 56–60
Rerkasem, B. & K. 236
resources 38
  biological 19
  local level management of 179–208
Rhoen Biosphere Reserve, Germany xiv, 227–32
Robertson, I. 33
Robinson, D. 169–71
Robson, C. 105–6
Roling, N. 201
rural development 6, 111–25
  and agrarian change 112–13
  case study 118–25
  importance of 111–12
  and poverty 115–18
  and protected area management 111–12, 118–25
  as social science practice 113–15

Sagarmatha National Park, Nepal 169–71
sampling 65–6
Sarin, M. 101, 189, 190, 191–2, 193, 199
Sattler, P. 17
Sayer, J. 212, 213
Scherl, L. 182
Scoones, I. 76
self-interested rationality 46
Shadish, W. 87
Simpson, T. 130
Smith, R.C. 138
Smyth, D. 127
social change 46
social impact assessment 60–3
social institutions *see* institutions, social
social sciences x, 31–47
  anthropology 33–7
  community consultation 101–8
  concepts 41–7
  definition 31–2
  economics 37–40
  institution building and strengthening 93–101
  methodologies and techniques 48–92
  and rural development 113–15
  sociology 33–7

social stratification 47
social transformation 47
socialisation 47
  agencies of 41
society 47
  and biodiversity 19–20
Society for the Promotion of Wastelands Development 189
sociological imagination 47
sociology 33–7
species diversity 16–18
Spradley, J. 51, 71, 73, 74
Stafford, C. 42
state
  decentralising power 97
  policy 35–6
stratification, social 47
Strong, M.F. 132
surveys 63–9
sustainability x, 180, 246
sustainable development 13
sustainable tourism 149–50
Suzuki, D. 132

Tanzania, East Usambara Mountains xiv, 213–14
Thailand, Mae Sa-Kog Ma Biosphere Reserve xiv, 236–8
theory-building 71
Thwaites, R. 218, 220
Tisdell, C. 38, 42
Tjamiwa, T. 144
tourism 147–8
  *see also* ecotourism
Toyne, P. 200
Tracey, P. vii
transformation, social 47
transition zone, of a biosphere 212
Traore, A. 226
triangulation 56

Uluru-Kata Tjuta National Park, Australia xiv, 139–45, 156
United Nations (UN)
  Commission on Human Rights 128, 131
  Conference on Environment and Development (UNCED) 3, 134, 216
  Environment Program (UNEP) 4

# INDEX

International Year for the World's Indigenous Peoples 131
*List of National Parks and Protected Areas* ix
Working Group on Indigenous Populations 127, 130, 134
United Nations Educational, Scientific and Cultural Organisation (UNESCO)
  Biosphere Conferences 209, 216–18, 219
  international protected areas 27
United States of America (USA)
  national parks 25
  rapid ethnographic assessment procedures (REAP) 53
  social impact assessment 60–3
unstructured interviews 70–1
Uphoff, N. 96

Valentine, P. 149
value, economic 43
  and tourism 157–8
values 47
van Willigen, J. 85, 86

Wadsworth, Y. 91
Wallis, A. 90–2, 99–101, 102
Walsh, R. 82
wealth ranking in agricultural communities, Zimbabwe xiv, 76–7
welfare economics 38–9, 42
Wells, M. 3, 164–6
West, P. 179
West Estonian Archipelago Biosphere Reserve, Estonia xiv, 232–3

Western, D. 150, 151–2
Williams, N. 138
Wiltshire, G. vii
Woenne-Green, S. 140, 144
Wolf, E. 129
Woodhill, J. 201
World Commission on Environment and Development 13, 133–4
World Conservation Monitoring Centre (WCMC) 28
World Council of Indigenous Peoples 131
world heritage sites 27
World Resources Institute 21, 23–4
World Wide Fund for Nature 185
Worldwatch Institute 128
Wright, P. 166, 168
Wright, R. 126, 131, 132
Wright, S. 37
Wynne, B. 75

Xilingol Biosphere Reserve, China xiv, 118–25, 155, 218–26
Xu, Z. 219

Yi Zhou Shu 25
Young, M. 155, 180

Zambia, safaris 156
Zhao, X. 218
Zimbabwe
  CAMPFIRE program xiv, 79, 156, 183–9, 206–7
  wealth ranking in agricultural communities xiv, 76–7

*Index compiled by Susan J. Ramsey*